MANUAL OF ANTISENSE METHODOLOGY

THE KLUWER SERIES

PERSPECTIVES IN ANTISENSE SCIENCE

Why a series of new volumes on antisense oligonucleotides? Because of the enduring fascination with the antisense biotechnology, which in theory gives the scientific and therapeutically-oriented communities the ability to sequence-specifically inhibit protein translation and hence the expression of genes.

At any rate, that's the theory. In practice, as is well known, the application of that theory to solve real biological problems presents a series of tortuous new problems, many of which are just now beginning to be understood, and hopefully resolved. Nevertheless, the progress in antisense biotechnology in the past few years alone has been impressive, and indeed the first antisense drug, fomivirsen (for cytomegalovirus retinitis: Isis, Carlsbad, CA) has recently been approved by the FDA.

The reader will find no dearth of well designed experiments in these volumes that demonstrate, to the best of current technology, sequence specific inhibition of gene expression. Much effort has also been expended by many authors in critical analysis of their results, a process always necessary for proper interpretation of data derived from antisense experiments.

There is little doubt that the coming years will witness further improvements and refinements in this dynamic technology, driven not only by the power of the idea, but also by the necessity generated by the sequencing of the human genome. These volumes therefore represent only the beginning of the harnessing of this impressive potential.

It was an honor for me to be asked to be series editor for these volumes, none the least because it gave me a chance to extensively interact with an excellent series of individual volume editors, who, at the time of this writing, included Stefan Endres, Peg McCarthy, Claude Malvy and LeRoy Rabbani. The results are mostly a product of their efforts, and of course even more so those of large number of authors. Finally, on behalf of all contributors, I would like to thank Charles Schmieg of Kluwer, who conceived of and drove this project, and without whom this collection would not exist.

C. A. Stein, Series Editor

Recently Published Book in the Series

Margaret M. McCarthy:
Modulating Gene Expression By Antisense Oligonucleotides
To Understand Neural Functioning
Claude Malvy, Annick Harel-Bellan, and Linda L. Pritchard:
Triple Helix Forming Oligonucleotides
LeRoy E. Rabbani, M.D.:
Applications Of Antisense Therapies To Restenosis

MANUAL OF ANTISENSE METHODOLOGY

edited by

Gunther Hartmann
and
Stefan Endres

University of Munich

PERSPECTIVES IN ANTISENSE SCIENCE
Series Editor: C. A. Stein
The College of Physicians & Surgeons, Columbia University

KLUWER ACADEMIC PUBLISHERS
Boston / Dordrecht / London

Distributors for North, Central and South America:
Kluwer Academic Publishers
101 Philip Drive
Assinippi Park
Norwell, Massachusetts 02061 USA
Telephone (781) 871-6600
Fax (781) 871-6528
E-Mail <kluwer@wkap.com>

QP
623
.5
.A58
M36
1999

Distributors for all other countries:
Kluwer Academic Publishers Group
Distribution Centre
Post Office Box 322
3300 AH Dordrecht, THE NETHERLANDS
Telephone 31 78 6392 392
Fax 31 78 6546 474
E-Mail <services@wkap.nl>

 Electronic Services <http://www.wkap.nl>

Library of Congress Cataloging-in-Publication Data

Manual of antisense methodology / edited by Gunther Hartmann and
    Stefan Endres.
        p. cm.--(Perspectives in antisense science)
    Includes bibliographical references and index.
    ISBN 0-7923-8539-X (alk. paper)
        1. Antisense nucleic acids--Research--Methodology. I. Hartmann,
    Gunther, 1966- . II. Endres, Stefan, 1957- . III. Series.
    QP623.5.A58M36   1999
    572.8'4--dc21   99-24715
                                                           CIP

Printed on acid-free paper.

Printed in the United States of America

TABLE OF CONTENTS

PREFACE

Antisense oligonucleotides are short single stranded synthetic nucleic acids designed to block the formation of the target protein by complementary binding to the corresponding target RNA. As simple the original idea was, as difficult has been its successful transfer into well-controlled experimental settings. In the past few years, the antisense approach became a scientifically well-based technology, which holds substantial therapeutic promise. This development is underlined by the first antisense drug on the market, fomivirsen, which has been recently approved by the Food and Drug Administration for the therapy of cytomegaly virus (CMV) retinitis in patients suffering from the acquired immunodeficiency syndrome (AIDS).

The efforts in the antisense field are accompanied by a tremendous progress towards the understanding of the behaviour of nucleic acids in biologic systems, whose solely function was originally thought to be the blueprint of life. Exciting non-antisense properties of oligonucleotides have been discovered, broadening our knowledge of biologic actions of nucleic acids. Nonantisense oligonucleotides containing a defined sequence motif (the dinucleotide CpG with specific flanking bases) are under investigation as therapeutics as well as antisense oligonucleotides. Furthermore, so-called therapeutic oligonucleotides are designed to combine an intended antisense with an useful non-antisense effect. The overall biological effect of any oligonucleotide is the sum of antisense and non-antisense actions. Dependent on the experimental setting, e. g. in cell-free systems an almost pure antisense-mediated mechanism may be achieved. The more complex a biological system is, the more difficult is the discrimination between antisense and non-antisense effects.

Besides the use as drugs, antisense technology may be critical to the development of functional genomics. This becomes increasingly important as the human genome project uncovers the sequence of numerous new genes with unknown function. Functional analysis of new genes with antisense oligonucleotides could gain tremendous impact for the identification of pharmacologic target. Once identified, the expression of these targets can be influenced therapeutically either by conventional or by gene-based pharmacologic strategies.

We have organised this book beginning with the biochemical background of the antisense strategy (Part I), continuing with different aspects of the application of antisense oligonucleotides in cell culture (Part II), and finally providing information on the use of this strategy in animal models and in clinical studies (Part III).

Oligonucleotide synthesis, the basis of antisense technology, is describes in chapter 1. Chapter 2 introduces methods which allow a preselection of antisense sequences. These methods are based on the physical interaction of oligonucleotides with their target RNA and on the use of cell-free translational systems. Even if the actual biological potency of an antisense oligonucleotides has to be checked empirically in cell culture, the proposed methods narrow the range of sequences to be tested. Spontaneous oligonucleotide uptake especially in primary cells is often low and needs to be enhanced by transfection with cationic lipids. Quantification of oligonucleotide uptake, which is described in chapter 3 for human leukocytes, allows to optimize transfection protocols for each cell type used. Targeted uptake of oligonucleotides in hepatocytes is described in chapter 4. The next logical step developing a valuable antisense model is to examine the system for non-antisense effects of the oligonucleotides used. Chapter 5 summarizes the current knowledge of nonantisense effects and provides advices how to avoid them. Chapter 6 provides state of the art guidelines and protocols for antisense experiments *in vitro,* which especially address common pitfalls. Chapter 7 and 8 present examples of well controlled antisense experiments with the pro-inflammatory cytokine tumor necrosis factor-α (TNF) and the receptor for the hormone insuline-like growth factor-1 (IGF-1 receptor) as targets. The protocols can be followed step by step. For development of therapeutic antisense oligonucleotides, animal models have to confirm the *in vivo* efficacy of a compound active in cell culture. Chapter 9 demonstrates well established animal models with the central nervous system as target tissue. Chapter 10 gives future perspectives by presenting the use of lipid-based carriers to improve pharmacokinetic properties of systemically delivered antisense oligonucleotides. Finally, general guidelines and requirements for the initiation of clinical studies are summarized in chapter 11.

Some essential issues are covered by several chapters independently. Different modifications of oligonucleotides are discussed in chapters 6 and 9. Detailed protocols for the improvement of oligonucleotide uptake are provided in chapters 3, 4, 6 and 10. Control experiments are specifically discussed in chapters 2, 5, and 6.

This book focuses on the application of antisense oligonucleotides and does not include other strategies applying an antisense mechanism, such as ribozymes or triple helix approaches. The intention of this book is not to be an exhaustive descriptive review of all opinions in the field of antisense oligonucleotides. Providing a state of the art knowledge, the book is designed to be a useful guide to successful antisense experiments with emphasis on the avoidance of common experimental pitfalls. We hope that this book will benefit researchers as well as clinicians, and will form a basis for further optimizing the methodology,

Sincere thanks are due to Felix Kratzer, Oliver Blank and Sabine Pflughaupt, who have been particularly helpful in formatting and proof reading of the book. We are very grateful to the editor of this book series, Cy Stein, and to Arthur Krieg for advice and guidance. Furthermore, it was a constant delight to work with several people in the Division of Clinical Pharmacology at the University of Munich, who have provided scientific inspiration and humor, and thus were also indirect contributors to this volume, These include Katharina Tschöp, Andreas Eigler, Ulrich Hacker, Jochen Moeller, Britta Siegmund, Christoph Bidlingmaier, Evelyn Tuma, and Verena Matschke. Finally, it is a pleasure to be the mentor for Anne Krug, Bernd Jahrsdoerfer and Steffanie Thompsom, who have contributed to our work with antisense oligonucleotides.

Munich
Iowa City
September 1998

Stefan Endres
Gunther Hartmann

Contributors

Mark Andrade (Chapter 1)
ISIS Pharmaceutical, Inc.
2292 Faraday Avenue
Carlsbad, CA 92008
USA

Martin Bidlingmaier (Chapter 3)
Division of Neuroendocrinology, Medizinische Klinik
Klinikum Innenstadt of the Ludwig Maximilians University
Ziemssenstr. 1
80336 München
Germany

Hubert E. Blum (Chapter 4)
Department of Medicine II
University of Freiburg
Hugstetter Str. 55
79106 Freiburg
Germany

Heinrich Brinkmeier (Chapter 2)
Department of General Physiology
University of Ulm
Albert-Einstein-Allee 11
89081 Ulm
Germany

Douglas L. Cole (Chapter 1)
ISIS Pharmaceutical, Inc.
2292 Faraday Avenue
Carlsbad, CA 92008
USA

Finbarr E. Cotter (Chapter 11)
Molecular Haematology Unit, Institute of Child Health
30 Guilford Street
London WC1N 1EH
UK

Rantjit R. Deshmukh (Chapter 1)
ISIS Pharmaceutical, Inc.
2292 Faraday Avenue
Carlsbad, CA 92008

USA

Dean Fennell (Chapter 11)
Molecular Haematology Unit
Institute of Child Health
30 Guilford Street
London WC1N 1EH
UK

W. Michael Flanagan (Chapter 6)
Department of Cell Biology
Gilead Sciences
333 Lakeside Drive
Foster City, CA 94404
USA

Gunther Hartmann (Chapter 3)
Division of Clinical Pharmacology, Medizinische Klinik
Klinikum Innenstadt of the Ludwig Maximilians University
Ziemssenstr. 1
80336 Munich
Germany

Lars Holmberg (Chapter 1)
Amersham Pharmacia Biotech
Milwaukee, WI 53202
USA

Michael J. Hope (Chapter 10)
Inex Pharmaceuticals Corporation
100-8900 Glenlyon Parkway,
Burnaby, British Columbia, V5J 5J8
Canada

Sandra K. Klimuk (Chapter 10)
Inex Pharmaceuticals Corporation
100-8900 Glenlyon Parkway,
Burnaby, British Columbia, V5J 5J8
Canada

Lester Kobzik (Chapter 7)
Harvard School of Public Health
665 Huntington Avenue
Building II - Room 231
Boston, MA 02115
USA

Arthur M. Krieg (Chapter 5)
Department of Internal Medicine
University of Iowa
540 EMRB
Iowa City, IA 52242
USA

Anne Krug (Chapter 3)
Division of Clinical Pharmacology
Medizinische Klinik
Klinikum Innenstadt of the Ludwig Maximilians University
Ziemssenstr. 1
80336 Munich
Germany

Jerzy Madon (Chapter 4)
Department of Medicine
University Hospital
Zürich, Switzerland

Deborah C. Mash (Chapter 9)
Molecular and Cellular Pharmacology
School of Medicine
P.O. Box 016960 D 4-5
Miami, Florida 33101

Darius Moradpour (Chapter 4)
Department of Medicine II
University of Freiburg
Hugstetter Str. 55
79106 Freiburg
Germany

Wolf-Bernhard Offensperger (Chapter 4)
Department of Medicine II
University of Freiburg
Hugstetter Str. 55
79106 Freiburg
Germany

Christoph Probst (Chapter 9)
Max-Planck-Institute for Neurobiology
Kraepelin Str. 2-10
80804 Munich
Germany

Mariana Resnicoff (Chapter 8)
Kimmel Cancer Center
223. S 10thStreet, 606 BLSB
Philadelphia, PA 19107, USA

Yogesh S. Sanghvi (Chapter 1)
ISIS Pharmaceutical, Inc.
2292 Faraday Avenue
Carlsbad, CA 92008
USA

Bernhard Schu (Chapter 2)
Department of General Physiology
University of Ulm
Albert-Einstein-Allee 11
89081 Ulm
Germany

Anthony N. Scozzari (Chapter 1)
ISIS Pharmaceutical, Inc.
2292 Faraday Avenue
Carlsbad, CA 92008
USA

Sean C. Semple (Chapter 10)
Inex Pharmaceuticals Corporation
100-8900 Glenlyon Parkway,
Burnaby, British Columbia, V5J 5J8
Canada

T. Skutella (Chapter 9)
Humboldt University
Institute of Anatomy
Berlin, Germany

Rainer Spanagel (Chapter 9)
Max-Planck-Institute for Psychiatry
Drug Abuse Group
Kraepelin Str. 2-10
80804 München
Germany

Margaret Taylor (Chapter 7)
Harvard School of Public Health
665 Huntington Avenue
Building II - Room 231
Boston, MA 02115
USA

Christian Thoma (Chapter 4)
Department of Medicine II
University of Freiburg
Hugstetter Str. 55
79106 Freiburg
Germany

Murray S. Webb (Chapter 10)
Inex Pharmaceuticals Corporation
100-8900 Glenlyon Parkway,
Burnaby, British Columbia, V5J 5J8
Canada

Fritz von Weizsäcker (Chapter 4)
Department of Medicine II
University of Freiburg
Hugstetter Str. 55
79106 Freiburg
Germany

Part I:
The Chemistry of Antisense Oligonucleotides

1 Chemical Synthesis and Purification of Phosphorothioate Antisense Oligonucleotides

Yogesh S. Sanghvi, Mark Andrade,
Rantjit R. Deshmukh, Lars Holmberg,
Anthony N. Scozzari and Douglas L. Cole

ISIS Pharmaceuticals Inc.
Developments Chemistry Department
Carlsbad, CA, USA

1.1 INTRODUCTION

Over a dozen antisense oligonucleotide drugs are undergoing human clinical trials for the treatment of viral infections, cancers, and a range of inflammatory disorders (Table 1). One of these was recently the first antisense oligonucleotide to demonstrate clinical safety and efficacy in pivotal Phase III clinical trials, in this case for the treatment of cytomegalovirus retinitis (Sanghvi et al., 1998). A dozen more antisense oligonucleotides have demonstrated pre-clinical efficacy (Crooke, 1998) and are under consideration for clinical development. In addition, use of antisense gene expression modulation to produce well-defined pharmacological effects is now a routine procedure (Akhtar et al., 1997). Automation of synthesis and ready access to required raw materials are two key reasons for the tumultuous growth in this area of research and development. Methods that allow preparation of a large number of pure oligonucleotides at reasonable cost expedite not only antisense drug discovery, but also open the door to manufacture of these drugs for market and the ultimate goal of delivery to patients. This chapter focuses on advances made in oligonucleotide process chemistry, the introduction and use of new reagents, and purification and analysis of antisense oligonucleotides.

Table 1 List of phosphorothioate oligonucleotides undergoing human clinical trials

Oligo No. or Name (Sponsor)	Sequence (5'-3')	Target (Disease)	Route of Administration	Status (Phase)
Fomivirsen (Isis)	gcgtttgctcttcttcttgcg	IE-2 (CMV retinitis)	Intravitreal	FDA-apprvd.
2302 (Isis)	gcccaagctggcatccgtca	3'-UTR / ICAM-1 (Crohn's disease psoriasis rheumatoid Arthritis ulcerative colitis renal allograft)	Intravenous	II a/b
3521/CPG 64128A (Isis/Novartis)	gttctcgctggtgagtttca	3'-UTR / PKC-α (Ovarian cancer) and variety of solid tumors	Intravenous	II a
5132/CPG 69846A (Isis/Novartis)	tcccgcctgtgacatgcatt	c-RAF kinase (Breast, prostate, colon, brain and ovarian cancer)	Intravenous	I/II
2503 (Isis)	tccgtcatcgctcctcaggg	Ha-ras oncogene (Variety of solid tumors)	Intravenous	I
G3139 (Genta)	tctcccagcgtgcgccat	bcl-2 / proto-oncogene (Non-Hodgkin's lymphoma)	Subcutaneous infusion	I/II a
LR3280 (Lynx)	aacgttgaggggcat	c-myc / proto-oncogene (Stent restenosis)	Intracoronary	I/II
LR3001 (Lynx)	tatgctgtgccggggtcttcg ggc	c-myb / proto-oncogene (Chronic myeloid leukemia)	Purged bone marrow	II
LR4437 (Lynx)	ggaccctcctccggagcc	IGF-IR (Ex-vivo tumor cells)	Intra-abdominal implant	I
GEM-132* (Hybridon)	<u>ugg</u> ggcttaccttgc <u>gaaca</u>	Intron-exon / UL36/27 (CMV retinitis)	Intravitreal	I/II
GEM-92* (Hybridon)	<u>ucgc</u> acccatctctctc <u>cuuc</u>	Gag/HIV-1 (AIDS)	Intravenous	I
GEM-231* (Hybridon)	<u>gcgu</u> gcctcctcac <u>uggc</u>	pka-1 (Refractory solid tumors)	Intravenous	I
GPI-2A† (Novopharm)	g'gttc'ttttg'g'tcc'ttg'tc't	Gag / HIV-1 (AIDS)	Liposome formulation Intravenous	I
13312⁺ (Isis)	<u>gc'gttt</u> gc'tc'ttc't <u>tc'ttgcg</u>	IE-2 (CMV retinitis)	Intravitreal	I

* The underlined bases in GEM-132, 92 and 231 are 2'-OMe sugar modifications.

† In GPI-2A, there are seven PS linkages represented by ' and the rest of the oligo is a phosphodiester.

+ In 13312, underlined bases are 2'-O(CH$_2$)$_2$OCH$_3$ sugar modifications and all U and C residues are 5-methyl substituted.

Interestingly, all first generation antisense oligonucleotide belong to one DNA analogue class, popularly known as phosphorothioates (PS) (Cohen, 1993). This chapter is confined to discussing recent trends in the chemical synthesis, purification and analysis of the PS oligonucleotides. The clinical success of this antisense oligonucleotide class presents a great advantage to the process chemist, in that once an investment in manufacturing process technology has been made for one PS drug, the same process will be applicable to the others, thus reducing overall cost of developing antisense therapies. We anticipate (Holmberg, 1997) that market demand for antisense treatment of inflammation and cancers may require a metric ton of PS drugs. With this requirement in mind, what follows is a discussion and update of current status and what the future may hold for manufacture of oligonucleotide drugs at large scale.

1.2 MATERIALS

The list of equipment and chemicals herein reflects general preferences, but does limit the scope of oligonucleotide manufacture. Where materials are readily available, we provide names of representative commercial suppliers.

Equipment

Equipment discussed below has been used routinely at Isis Pharmaceuticals Inc. (Isis) to manufacture antisense oligonucleotide in hundred-gram to multi-kilogram quantities.

Synthesis

The Pharmacia OligoPilot II (OP II) synthesizer is a fully automated computer-controlled flow-through reactor system for 100 µmol to 4 mmol scale synthesis of antisense oligonucleotide. Oligonucleotide synthesis protocols developed on the OP II can be scaled up in a linear manner for use on the Pharmacia OligoProcess synthesizer at run-scales from 10 mmol to 180 mmol. In addition, a large-scale synthesizer (OligoMax) is under development by Pharmacia and Isis that will enable synthesis between 0.2 mol and 2 mole scale. The completed OligoProcess will have metric ton annual output capacity. Other pilot scale synthesizers are the (i) Millipore 8800 [2]; (ii) Hybridon 601 [3]; and (iii) ABI 390Z [4].

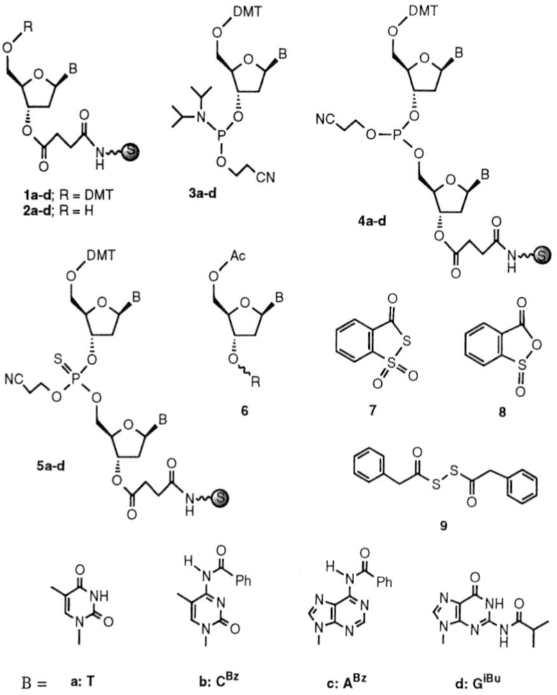

Figure 1: Synthesis Reagents

Purification

Most pilot-scale purifications (1-10 g) were carried out on a PerSeptive Biosystems BioCad 60 HPLC [5] connected to an SF-2120 fraction collector [6]. A PerSeptive Biosystems Vision HPLC [5] was also used for automated analysis of collected fractions from preparative runs. Clinical material was purified on a Biotage KiloPrep 100 LC [7] and a radially compressed Waters BondaPak HC$_{18}$-HA cartridge [8]. Purified clinical material was precipitated and collected using a Carr [9] continuous flow-through centrifuge. Residual OVIs were removed in a Leybold moving-shelf freeze-drier [10], providing a convenient and stable storage form for kilogram quantities of antisense oligonucleotide.

Analysis

Mass spectra were acquired using an LCQ quadrupole ion trap mass spectrometer equipped with an electrospray ionization source [11]. Desalted oligonucleotides were analyzed by Quantitative Capillary Gel Electrophoresis using a Beckman P/ACE 5010 capillary electrophoresis instrument [12]. Analytical HPLC was carried out on a Waters chromatographic system with 717 auto sampler, 600E controller, 991 photodiode array detector and Millennium 2.01 operating software [8]. Sequencing of oligonucleotides was carried out by MALDI-TOF mass spectrometry [13]. ^{31}P NMR spectra were recorded on a Varian Unity 400 MHz instrument [14]. DNA duplex melting temperatures were measured on a Gilford Response II 260 spectrophotometer [15].

Chemicals and Reagents

Key chemicals and reagents required for synthesis of PS oligonucleotides are as followed (Figure 1). Solid supports: **1a** T, **1b** dC^{Bz}, **1c** dC^{Bz}, **1d** dG^{iBu}; **1a-d** protected 2'-deoxynucleosides are anchored to commercial polystyrene support HL-30 [1] via a succinyl linkage. Phosphoramidites: **3a** T, **3b** dC^{Bz}, **3c** dA^{Bz}, **3d** dG^{iBu}; Amidites **3a-d** are protected with conventional (Beaucage et al., 1996) protecting groups. Deblock: A solution of dichloroacetic acid (DCA) in dichloromethane (DCM). Activator: A solution of 1*H*-tetrazole in anhydrous acetonitrile (ACN) is employed for coupling reactions. Sulfurization: A solution of Beaucage reagent in anhydrous ACN is used. Capping: Cap A solution is NMI in pyridine and anhydrous ACN and Cap B solution is Ac$_2$O in anhydrous ACN. Cleavage and deprotection: Concentrated NH$_4$OH solution is used as such.

Purification: Standard purification requires the following items. Water: USP Purified Water is used for chromatography. Buffers: A solution of dilute NaOAc was used as a buffer for RP chromatography. Two buffers were employed for gradient IE chromatography. Buffer A: 50 mM NaOH and buffer B: 50 mM NaOH containing NaCl. Organic: Reagent grade MeOH was used for RP column chromatography. Industrial grade 95% EtOH was used for product precipitation.

Analysis: A variety of reagents are required in small amounts. All of these reagents are listed in the appropriate publications (see Table 2).

1.3 METHODS

The four typical steps in manufacture of PS oligonucleotide drugs are: (i) chemical synthesis, (ii) purification, (iii) any required desalting, and (iv) control analysis, shown in Figure 2. Experimental details of each step are elaborated on in this section.

Table 2. Typical tests required for phosphorothioate antisense oligonucleotide drug control

Purpose	Method/Equipment	Reference
MW determination	ES/MS	Nordhoff et al., 1996
Retention time	IEX HPLC	Chapter 12, Ion Exchange Chromatography, 1997
P=S vs. P=O ratio	^{31}P NMR	Jaroszewski et al., 1994
Measure duplex T_m	UV	Breslauer, 1994
Sequencing	MALDI-TOF MS	Schuette et al., 1995
Area-% impurity profile	IEX-HPLC	Srivatsa, in IBC meeting, 1997
Assay vs. external reference	CGE	Srivatsa et al., 1994
Organic volatile impurities	Capillary GC	Analytical Gas Chromatography, 1997
Heavy metals	ICP-MS	Loorab et al., 1998
Residual buffer salts	CZE	Chen et al., 1997
Endotoxins	LAL	Endotoxins and their detection, 1980

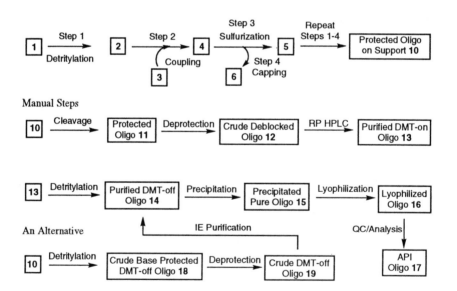

Figure 2: Automated Solid-Phase Synthesis

Chemical Synthesis

The chemical assembly of antisense oligonucleotide takes place in two stages: first, automated solid-phase synthesis, and second, a manual cleavage and deprotection step. This two-step process is the fastest and best method known at the present time (Holmberg, 1998).

In order to appreciate the current state-of-the-art in solid-phase oligonucleotide synthesis, a short historical summary of the evolution of large-scale automated synthesis may be useful. The first serious solid-phase scale-up attempt at Isis began in 1989 using a bank of stirred-bed Millipore 8800 synthesizers that could make gram quantities of PS oligonucleotides per cycle. Large consumption of amidite monomers and, solvents and the low loading of solid supports in use at the time restricted scale-up with 8800 synthesizer. Isis therefore determined to investigate potentially more efficient packed-bed reactor designs. In close collaboration with Large Scale Biology (Gaithersburg, Maryland), we designed and built the first large-scale packed-bed oligonucleotide synthesis reactor, the LSB PCOS-2, based on a centrifugal liquid delivery concept. The excellent performance of the PCOS-2 confirmed our hypotheses that some solid-supports were rugged enough for packed-bed use, that packed reactor beds were more efficient in solvent and amidite consumption, and very importantly, that sulfurization of phosphite triester intermediates to phosphorothioates in a flow-through system did not lead to increased levels of phosphodiesters. This success paved the way for the design of larger synthesizers using conventional chromatography columns as reactors. In 1993, Isis and Pharmacia partnered to design, build and validate an oligonucleotide synthesizer based on a fixed-bed low aspect ratio chromatography column. During the last five years, a series of three such synthesizers of increasing capacity (the Pharmacia OligoPilot I, the OligoPilot II, and the OligoProcess) were built and successfully tested. More importantly, a straightforward and linear scale-up from less than 1 mmol to 150 mmol was accomplished as each Pharmacia synthesizer was put to routine use at its full design scale. Currently, the same chemists and engineers are designing a 2 mole synthesizer (dubbed the "OligoMax") to produce over 6 kilo of pure phosphorothioate per day, equivalent to an annual output on the order of one metric ton of antisense oligonucleotides.

The focus of this chapter is production of clinical and market supplies of PS oligonucleotides, so only current large-scale (1 to 150 mmol) manufacture will be reviewed further. Excellent review articles and books exist, however, on the synthesis, (Protocols for Oligonucleotides and Analogs, 1993) purification (Deshmukh et al., 1998), and analysis (Protocols for Oligonucleotide Conjugates, 1994) of oligonucleotides at scales below the 1 mmol level.

Automated Solid-Phase Synthesis

Assembly of the protected antisense oligonucleotide is carried out by first packing a column with solid support previously derivatized with the desired 3'-nucleoside **1** then passing reagents and solvents through in a predetermined order. Attachment of each nucleotide residue to the growing chain support requires four reactions. All reactions are sequential and reagent and solvent delivery are computer-controlled. Each of the four steps is associated with a number of solvent washes to remove byproducts and unreacted reagents from the column. The purpose of multiple dry solvent wash cycles is to keep the support-bound antisense oligonucleotide under anhydrous conditions, a key requirement for efficient coupling of nucleoside phosphoramidites. Synthesis of a 20-mer PS oligonucleotide (Table 3), including at 150 mmol scale, can be completed in less than 12 hours without active operator intervention[16]. The experience of scaling-up from 10 mmol to 180 mmol of oligonucleotide can be summarized as follows. With the scale-up, purity of the crude product is maintained or in some cases improved which in turn may lower the cost of manufacturing on large-scale. The cycle time has been kept under 30 minutes throughout the scale-up allowing to complete the campaign in a normal work day. The consumption of solvents and some of the ancillary reagents went down on a mmol basis with the scale-up. For example, the volume of deblock solution used on a 100 mmol scale is almost 45% less per mmol compared to the 10 mmol synthesis (Table 3). Furthermore, with the scale-up relative cost of the labor has been reduced lowering the cost of production of antisense oligonucleotide drugs. Our limited experience (5 campaigns) at 150 mmol scale is in line with the expected purity, shorter cycle time, and lower consumption of solvents and reagents.

Table 3. Oligo Process: Synthesis Campaign Details

Reagent/Solvent Consumption (per cycle in liter)	Scale (mmol)	
	10	100
ACN	6	45
Deblock solution (DCM)	3.5	20
Beaucage Reagent	0.16	1.6
1H-Tetrazole	0.4	1.5
Amidite	0.15	0.75
Capping A solution	0.1	1.0
Capping B solution	0.1	1.0
Other Details		
mmol of crude oligo obtained	7	70
% of n-mer in crude mixture	73%	73%
mmol of n-mer in crude	5	51
Pure n-mer obtained after purification	45%	45%
Pure precipitated n-mer (mmol)	4	44
Weight of pure product	30g	300g
% of full length product isolated	86%	85%
% yield based on support-loading	44%	44%

Step 1. Detritylation of **1** (cleavage of DMT-group) is accomplished with DCA in DCM, which releases DMT cation (bright red-orange)[17] leaving an unprotected 5'-OH group on the sugar residue bound to the support, such as **2**. The progress of the detritylation step is monitored by on-line UV detection, allowing the pumping of DCA/DCM solution through the column until no more cation is detected. Detritylation under acidic conditions is known to be reversible. Therefore, in order to drive this reaction to completion, we (Ravikumar et al., 1995) have employed a trityl cation scavenger that can aid efficiency of solid-phase synthesis. The column is then washed with anhydrous ACN. However, the reported toxicity (Hazardous Chemical Desk Reference, 1997) of DCM forced us to search for alternatives. This search led to the discovery (Krotz et al., submitted) of toluene as an alternative solvent for deblock solutions.

Step 2. The next reaction involves condensation of an appropriate monomeric amidite **3** with the 5'-OH group generated in Step 1. The 1H-tetrazole serves as an activator[14] in this step by converting an amidite **3** into a reactive P(III) species that readily couples providing an internucleosidic phosphite triester linkage, as shown in **4**. Notably, this reaction always goes in yields near 99%, resulting in excellent full-length product yields overall. Furthermore, this reaction step has been successfully optimized to use as little as 1.5 molar excess of the amidite **3** on OP II and

OligoProcess synthesizers. After the coupling step, the support is washed again with ACN to remove excess reagents and byproducts.

Recently, 4,5-dicyanoimidazole (DCI) has been reported (Vargeese at al., 1998) as an improved and safer reagent than 1*H*-tetrazole as coupling activator. Preliminary results indicate that DCI may be a viable activator for large-scale synthesis if tetrazole handling or tetrazole-contaminated solvent recovery become safety issues.

Step 3. Next, the newly formed P(III) phosphite triester **4** is transformed to a P(V) phosphorothioate linkage **5** by sulfurization with 3*H*-1,2-benzodithiol-3-one 1,1-dioxide **7** (Boal et al., 1996). The conversion of **4** to **5** is nearly 100% efficient. However, a small percentage of P(III) species may remain unreacted in any given sulfurization step during elongation of the antisense oligonucleotide, which may result in the formation of short sequences (Reddy, 1995). Though **7** is widely used for sulfurization, it does have limitations. First, it can contribute over 20% to total cost of raw materials for antisense oligonucleotide manufacture. Second, byproduct **8** formed during use of **7** and is known to produce diester (P=O) linkages. Efforts to discover alternative reagents are in progress and phenylacetyl disulfide (Cheruvallath et al., in preparation) **9**, for example, shows great potential for large-scale PS antisense oligonucleotide synthesis (20). The step after sulfurization is washing the column with ACN.

Step 4. Capping protects any unreacted 5'-OH groups from undergoing subsequent coupling steps, giving compound **6**. This reaction is carried out with a mixture of NMI and Ac$_2$O in ACN. For safety reasons [21], it has recently been recommended that 2,4,6-collidine should be used in Cap B solution with 20% Ac$_2$O in ACN. As before, the support is washed with ACN to remove unreacted reagents before initiating the next reaction. Again, efforts have been made to develop (Zhang et al., 1996) improved alternative reagents for capping but none are presently in routine use at large-scale.

Manual Steps: Cleavage and Deprotection

On completion of the automated synthesis, the oligonucleotide **10** remains attached to the support with all nitrogen- and oxygen-protecting groups intact. The support is then typically removed from the column and heated in NH$_4$OH for several hours. Base treatment results in cleavage of the succinyl ester linkage and all base and backbone protecting groups (except DMT) in a single step to furnish **12**. Although this one-pot NH$_4$OH treatment works reasonably well, there have been promising reports of the use of alternate reagents, such as gaseous ammonia (Boal et al., 1996) and methylamine (Reddy et al., 1995). Analysis of PS oligonucleotides deprotected in NH$_4$OH often reveals some formation of phosphodiester linkages. Notably, this formal sulfur loss is completely avoided (Reese et al., 1997) when deblocking is carried out in the presence of 2-mercaptoethanol. During solid-supported synthesis, it is very difficult to estimate how much sulfur may have been lost in the deblocking step or during the sulfurization step due to incomplete sulfur transfer. Therefore, in the future, use of gaseous ammonia and/or 2-mercaptoethanol may be beneficial in improving the quality of PS oligonucleotides manufactured at large scale.

Purification

The importance of developing post-synthesis procedures for purification of antisense oligonucleotides sometimes has been overlooked, mainly for the following reasons. Early scale-up efforts were extrapolated from "gene-machine" practice, in which very large molar excesses of phosphoramidite monomers are used to drive coupling yields to near unity, resulting in a crude product which is nearly all full-length oligonucleotide.   While this practice can circumvent the need for chromatographic purification after deprotection, the relatively high cost of amidite monomers renders it impractical for large scale work, and the lack of a purification step makes it unacceptable for making medical grade antisense oligonucleotides. Instead, it has been preferable to optimize synthesis to minimize amidite consumption while maintaining high coupling efficiency, and to develop efficient chromatographic methods for removal of deletion sequence impurities from the full-length product.

All first-generation phosphorothioate antisense oligonucleotide are made by solid-support synthesis and impurity profiles are superficially similar among the class, it has been possible to assume that a single chromatographic purification protocol would serve for all oligomers. As a result, there has been no widespread research directed to developing separation methods for these compounds. It has generally proven necessary to develop specific purification protocols for each phosphorothioate oligomer. This work has thus been carried out almost exclusively during development of specific manufacturing processes.

Preparative chromatographic method development and validation require relatively large amounts of crude synthetic oligonucleotide, which until recently were costly and difficult to obtain.

Finally the decision to invest resources in separations research is necessarily cost-driven. The anticipated cost of purification is only about 5% of the cost of raw materials required to make antisense oligonucleotide. As noted above, however, the optimization of oligonucleotide synthesis for low monomer excess has made investment in purification process development cost effective. The emphasis in this area is increasing further with the emergence of chimeric antisense oligonucleotide and the associated use of more costly novel reagents and chemistries (Sanghvi et al., 1994) for their synthesis.

Composition of synthetic PS oligonucleotides

The crude PS oligonucleotide obtained from solid-support synthesis contains several classes of impurities in addition to the desired full-length oligonucleotide. There is a pool of deletion, or (n-m)-mer, impurities (n-1, n-2, n-3, etc.) arising from chain failure to elongate. Nucleotide deletions occur randomly and with essentially equal frequency at each position in the oligomer (Chen et al., in preparation). The most probable deletionmer length is n-1 and up to n-1 of these may be present in a given oligomer synthesized by monomer coupling. Through synthesis of all possible deletion sequences of a phosphorothioate oligonucleotide, it was shown that the (n-1)-mers have essentially identical chromatographic retention properties, regardless of their nearly identical base compositions, and identical capillary electrophoretic mobilities (Srivatsa et al., 1997). Crude synthetic PS oligonucleotides will also contain partial phosphodiester (PO) oligomers, generated by incomplete sulfurization, oxidation of intermediates, or during NH_4OH deblocking steps. The most probable partial phosphodiesters are the "monophosphodiester" [(PO)$_1$] species containing a single phosphate linkage. Other minor impurities are (n+1)-length species generated due to the acidic nature of the activator used in the coupling step (Krotz et al., 1997) and resultant "double coupling". The preparative purification of PS oligonucleotides is further complicated by broad elution profiles of these products relative to native DNA oligomers, caused by chirality (Lebedev et al., 1996) at the phosphorus centers. Nevertheless, the following three purification techniques are used frequently and effectively in our labs and facilities to remove and reduce levels of the above impurities in full-length PS oligonucleotides.

Reverse phase (RP) HPLC

RP-HPLC is a widely used technique for the purification and analysis of a variety of oligonucleotides (Warren et al., in Protocols for Oligonucleotide Conjugates, 1994). The general method readily resolves DMT-on oligonucleotides from DMT-off failure sequences in crude synthesis products, due to the hydrophobic nature of the DMT-on oligomers. For example, the crude DMT-on PS oligonucleotide **12**, after deprotection and cleavage from the support, is purified on a Waters HC-C18 HA radial compression column, selected for its flow characteristics, efficiency, durability, and high loading capacity. The column is eluted in gradient of water: MeOH in dilute NaOAc. In a typical purification, two major peaks are observed (Figure 3). The major peak (b) contains "trityl-off failure sequences" and protecting group residues and the second major peak (c) is the "trityl-on" product **13** along with minor quantities of trityl-on failure sequences (d in Figure 3). A late-eluting third peak containing non-product material is also seen. The elution profile is monitored by continuous UV absorption spectroscopy and the trityl-on **13** peak is collected.

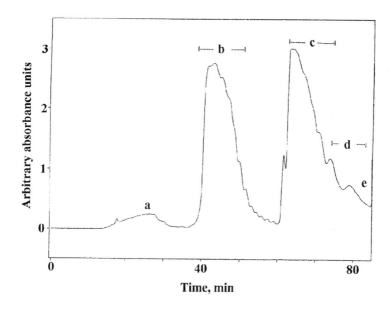

Figure 3: Preparative RP-HPLC of DMT-on antisense oligonucleotides
Media: Waters BondaPak HC$_{18}$HA; **System:** Biotage Kiloprep, with compression chamber; **Buffer A:** NaOAc in DI water; **Buffer B:** NaOAc in MeOH; **Gradient:** optimized gradients between 10% to 90% B; **Temperature:** Ambient; **Detection:** 260 nm; **Peak a, b:** DMT-off failures; **Peak c:** DMT-on full length product pool, **13; Peak d:** DMT-on failures; **Peak e:** wash peak.

The material is further processed through a detritylation step to furnish **14**, followed by precipitation to provide pure oligo **15**. Gratifyingly, large-scale radial compression columns are available in 20L bed volumes and similar buffers and gradients are readily scaled for larger scale manufacture of oligonucleotides. However, large-scale post-purification steps such as detritylation and precipitation are inconvenient and if repeated on smaller scale can add cost to the final bulk drug substance. For this reason, research has also been conducted into the best modes for utilizing ion exchange chromatography for oligonucleotide purification.

Tandem Purification: RPC followed by IEC

For high purity PS oligonucleotide requirements at large-scale, both RPC and IEC have been used, and can be used sequentially for double purification or "polishing". Although this method provides highly purified oligonucleotides, limitations remain, such as (i) reduced overall yield, (ii) extended handling of materials and associated potential for loss, and (iii) required optimization on multiple equipment. A double purification method is currently used for fomivirsen (Table 1). The crude DMT-on oligonucleotide **12** is first purified on a RP column as described in the previous section to give purified DMT-on oligo **13**. Detritylation of **13** followed by a second purification on an IEX column provides a final product of high full-length purity and low phosphodiester content, as expected based on the selectivities (Srivatsa et al., 1997) of these methods. Waters Protein Pak Q15 HR anion exchange media and an aqueous buffered sodium chloride gradient elution provide fomivirsen in greater than 71% yield with approximately 2% of (n-1)-mers and low phosphodiester content.

Ion Exchange (IEX) Chromatography

IEX liquid chromatography, as opposed to RP-HPLC, has been widely used for the preparative purification of biomolecules. Anion exchangers are the most frequently used preparative IEX media. For several reasons, we believe IEX is a superior choice for very large-scale separation of desired synthetic oligonucleotide products from their process-related impurities.

IEX LC (i) avoids use of organic solvents, (ii) can utilize less expensive equipment, (iii) accepts crude DMT-off oligonucleotide as feed, avoiding a post-RPC detritylation step, (iv) allows very high loading of crude antisense oligonucleotide, (v) easily removes DMT-cation from the product, and (vi) separates n-1-, n-2-, n+1-mers effectively, resulting in high product purity in a single step. The only IEX drawbacks are the required post-chromatography desalting step and the need to concentrate large volume aqueous fractions.

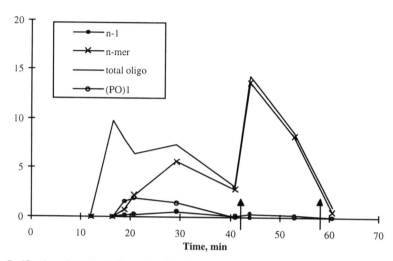

Figure 4: Purification of DMT-off oligonucleotides by single-column anion exchange chromatography. **Media:** POROS HQ 50 (PerSeptive BioSystems); **System:** BioCad 60 (PerSeptive BioSystems); **Buffer A:** aqueous NaOH; **Buffer B:** aqueous NaOH + aqueous NaCl; **Gradient:** Optimized step gradients; **Temperature:** ambient; **Detection:** 266 nm, 290 nm; **Run time:** 75 min; **Product yield:** 72%.

Several Isis antisense oligonucleotides (ISIS 2302, ISIS 3521, and ISIS 5132 in Table 1) have been purified by single-column IEX LC using sample self-displacement elution methods (Deshmukh et al., 1998; Deshmukh et al., 1997). General conditions for the purification of antisense oligonucleotide are as follows. The PS oligonucleotide is loaded onto a column packed with IEX media such as Q Hyper DF [22] or POROS HQ50 [5]. The saturation capacity of such media is 30-40 mg/ml. The elution gradient shown in Figure 4 requires 2 hours or less total run time. The n-1-mers and partial phosphodiesters elute early and tail toward the main peak. We believe n-1-mers and smaller oligomers are well displaced because of their weaker binding (ionic interaction) compared to full-length oligonucleotide. PO impurities are separated from the desired PS product based on the weaker affinity of phosphate diesters for anion exchange sites. To illustrate the power of the method, final purification of antisense oligonucleotide ISIS 5132 for reference standard use resulted in 99.4% full-length purity of the PS oligonucleotide as measured by analytical IEX-HPLC, with 0.6% $(PO)_1$, in 72% yield, beginning with reverse phase column eluate at 86.4% PS oligonucleotide and 8.5% of PO_1 content.

At high pH, IEX may be conducted without resolution and selectivity limitations stemming from the effect of charged heterocyclic bases or self-association in G-rich oligonucleotides. One such example is ISIS 5320 (TTGGGGTT) which is readily purified at strongly basic pH [23].

In summary, an array of chromatographic methods is available for purification of synthetic phosphorothioate oligonucleotides at large scales. RP-HPLC serves very well at multi-kilogram production scales. Compared to RP-HPLC and tandem RP-IEX purification, single-mode IEX LC offers a practical method for very large-scale manufacture due to its scalability through use of simple aqueous eluents, utility over a wide pH range, great resolving power, and the physicochemical robustness of the media.

Desalting

Excess buffer salts in intermediate **14** after RP-HPLC purification are removed in the precipitation step to obtain **15**. In some cases ultrafiltration (UF) may be used to further desalt the product. Two methods may be used to desalt oligonucleotide products after IEX purification: (i) RP-LC desalting and (ii) ultrafiltration. In the RP-LC method, purified antisense oligonucleotide **14** is loaded onto an RP column, the salts are removed with a DI water wash, and the desalted antisense oligonucleotide is eluted with an alcohol. This method is not well suited to very large-scale manufacture for reasons similar to those discussed in Section describing composition of synthetic PS oligonucleotides Ultrafiltration is a potentially superior method for very large-scale manufacture since it is widely used and appropriate equipment is readily available, and because the product is obtained as an aqueous solution. Problems with UF are (i) slow processing times, and (ii) difficulty in obtaining high concentration factors. Both problems stem from concentration polarization of oligonucleotides at the membrane surface. Therefore, proper process optimization is needed to provide an adequate amount of salt reduction and concentration for subsequent process steps. After desalting, the product may be freeze-dried to provide a dry powder storage form (**17**) or stored in the cold as a solution for subsequent use.

Analysis

Therapeutic use of synthetic antisense oligonucleotides requires not only efficient synthesis and purification, but also thorough in-process quality control, including the determination of process-related minor impurities. A self-consistent suite of analytical methods has been developed for complete characterization and control of antisense oligonucleotides (Srivatsa et al., 1997). Key raw materials analyzed and tightly controlled for quality are the nucleoside phosphoramidite monomers, solid supports, sulfurization reagents, and process solvents, typically available from commercial sources [1, 4, 5]. In-process separation-based control analyses by CGE and IEX-HPLC are carried out during purification and pooling of fractions. HPLC, CGE, and ^{31}P NMR determine the identity and sequence integrity of in-process and finished goods samples. The analytical methodologies required for control of therapeutic oligonucleotide final products have been detailed by FDA (Rao et al., 1993). Typical antisense oligonucleotide finished product tests are summarized in

Table 2. Since details of these analytical methods have been published in the literature (Nordhoff et al., 1996; Chapter 12 in "Ion-Exchange Chromatography", 1997; Jaroszewski at al., in Protocols for Oligonucleotide Conjugates, 1994; Breslauer, in Protocols for Oligonucleotide Conjugates, 1994; Wyrzykiewicz et al., 1994; Schuette et al., 1995; Srivatsa, 1997; Srivatsa et al., 1994; Analytical Gas Chromatography, 1997; ICP MS for Element Studies, 1998; Chen et al., 1997; Endotoxins and their Detection with Limulus Lysate Test, 1980), only recent developments are discussed herein.

In spite of the stability of phosphorothioate oligonucleotides to enzymolysis, a straightforward methodology for sequencing these nucleic acid analogs has been developed based on complete and mild oxidation to the corresponding full length native DNA structure, followed by either Maxam-Gilbert sequencing (Wyrzykiewicz et al., 1994) or MALDI-TOF mass.(Schuette et al., 1995). More recently, an alkylation technique (Polo et al., 1997) has allowed rapid sequence analysis of PS oligonucleotides without oxidation of the phosphorothioate linkages. A new interface procedure has been reported (Apffel et al., 1997) that combines HPLC-ESI/MS for the identification of oligonucleotides of similar composition but of varying sequence or conformation. An accurate and reliable method of quantifying sodium acetate in antisense oligonucleotide by CZE has been reported (Chen et al., 1997). The detection limit of this method is equivalent to 0.0017% of w/w of sodium acetate. A very interesting characterization of an n+ mer in an antisense oligonucleotide metabolite was demonstrated by combination of HPLC, CE and ES/MS methods (Cummins et al., 1997). Finally, the advent of various analytical techniques has had a significant impact on our understanding of process-related impurities and chemical and biological degradation products present in antisense oligonucleotide.

1.4 SUMMARY

During the past six years, the cost of synthetic phosphorothioate antisense oligonucleotide manufacture has been sharply reduced (~97%) while synthesis scale has been increased from about 200 micromoles per run to near 200 mmol per run (a net 1,000-fold scale-up). Antisense oligonucleotide purification by preparative reverse phase and ion exchange chromatographies has kept pace with these advances and scale increases, so synthesis and purification can be carried out in parallel at large scale. Environmentally objectionable process materials have been replaced. These advances were made by thoroughly optimizing the phosphoramidite coupling chemistries and phosphite triester sulfurization conditions established at the beginning of the period. Given the smooth 1,000-fold solid-phase synthesis scale-up achieved to date, it appears warranted to expect that a further 10-fold scale-up will likewise be successful. That success will position the industry for manufacture of the metric ton scale antisense oligonucleotides anticipated that will be required by the success of this new therapeutic technology.

1.5 REFERENCES

Akhtar S, Agrawal S. In vivo studies with antisense oligonucleotides. *TIPS* 1997,18 (and references cited therein).

Analytical Gas Chromatography, 2nd ed., 1997. Jennings W, Mittlefehldt E, Stremple P eds., Academic Press, San Diego, CA.

Apffel A, Chakel JA, Fischer S, Lichtenwalter K, Hancock WS. Analysis of oligonucleotides by HPLC-electrospray ionization mass spectroscopy. *Anal Chem* 1997;69:1320-25.

Beaucage SL, Caruthers MH. Deoxynucleoside phosphoramidites-a new class of key intermediates for deoxypolynucleotide synthesis. *Tetrahedron Lett* 1981;22:1859-62.

Beaucage SL, Caruthers MH. "The Chemical Synthesis of DNA/RNA", *Bioorganic Chemistry: Nucleic Acids* 1996;36-74. SM Hecht, ed. Oxford University Press.

Boal JH, Wilk A, Harindranath N, Max EE, Kempe T, Beaucage SL. Cleavage of oligodeoxyribonucleotides from controlled-pore glass supports and their rapid deprotection by gaseous amines. *Nucleic Acids Res* 1996;24:3115-17.

Breslauer KJ. Extracting thermodynamic data from equilibrium melting curves for oligonucleotide order-disclosure transitions. in *Protocols for Oligonucleotide Conjugates,* 1994;347-72. S Agrawal, ed. Humana Press.

Chapter 12 "Ion-Exchange Chromatography", *HPLC Columns: Theory, Technology and Practice*. 1997. 224-49. UD Neue, ed. Wiley-VCH.

Chen D, Klopchin P, Parson J, Srivatsa GS. Determination of sodium acetate in antisense oligonucleotides by capillary zone electrophoresis. *J Liq Chrom Rel Technol* 1997;20:1185-95.

Chen D, Yan Z, Cole DL, Srivatsa GS. Analysis of internal n-1 mer deletion sequences in synthetic oligonucleotide by hybridization to an immobilized probe array. Manuscript in preparation.

Cheruvallath ZS, et al. Investigations on the use of PADS in the synthesis of phosphorothioate oligonucleotides. Manuscript in preparation.

Cohen JS. "Phosphorothioate Oligodeoxynucleotides", *Antisense Research and Applications*. 1993;205-22. ST Crooke, B Lebleu, eds. CRC Press

Crooke ST. " Antisense Therapeutics" In *Biotechnology and Genetic Engineering Reviews*. Vol. 15, pp. 1998;121-57.

Cummins LL, Winniman M, Gaus HJ. Phosphorothioate oligonucleotide metabolism:characterization of the N+ mer by CE and HPLC ES/MS. *Bioorg Med Chem Lett* 1997;7:1225-30.

Deshmukh RR, Sanghvi YS. Recent trends in large-scale purification of antisense oligonucleotides. *IBC meeting,* 1997 October 28-29, San Diego, California

Deshmukh RR, Leitch WE II, Cole DL. Application of sample displacement techniques to the purification of synthetic oligonucleotides and nucleic acids: a mini-review with experimental results. *J Chromat* A 1998,806:77-92.

Endotoxins and their Detection with Limulus Amebocyte Lysate Test. Watson SW. Levin J, Novotsky, eds. Alan R Liss Inc., 1980.

Hazardous Chemical Desk Reference. 4th Edition. RJ Lewis, ed. New York: Van Nostrand Reinhold, 1997. 753-54.

Holmberg L. Scale-up of oligonucleotides to multiple tons per year. Large-scale oligonucleotide synthesis. *IBC meeting,* 1997 October 28-29, San Diego, California.

Holmberg L. Oligonucleotide manufacture under $50/gm. Antisense DNA and RNA-based therapeutics. *IBC Conference.* 1998 February 2-3, Coronado, California.

Jaroszewski JW, Roy S, Cohen JS. NMR studies of oligonucleotides. in *Protocols for Oligonucleotide Conjugates, 1994,* 301-17.

Krotz AH, Klopchin PG, Walker KL, Srivatsa GS, Cole DL, Ravikumar VT. On the formation of longmers in phosphorothioate oligodeoxyribonucleotide synthesis. *Tetrahedron Lett* 1997;38:3875-78.

Krotz AH., et al. DMT removal in organic solvent: efficient oligonucleotide synthesis without chlorinated solvent. Manuscript submitted.

Lebedev AV, Wickstrom E. The chirality problem in P-substituted oligonucleotides. *Perspectives in Drug Discovery and Design.* Vol. 4. 1996;17-40. GL Trainor, ed. ESCOM

Loorab GK, McKiernan JW, Caruso JA. ICP MS for element studies. *Mikrochim Acta* 1998;128:145-68.

Nordhoff E, Kirpekar F, Roepstorff P. Mass spectometry of nucleic acids. *Mass Spect Rev* 1996;15:67-138.

Polo LM, McCarley TD, Limbach PA. Chemical sequencing of phosphorothioate oligonucleotides using MALDI-TOFMS. *Anal Chem* 1997;69:1107-12.

Protocols for Oligonucleotides and Analogs. Chapters 2 (Christodoulou, C. Oligonucleotide Synthesis: Phosphotriester Approach), 3 (Beaucage S. Oligodeoxyribonucleotide Synthesis: Phosphoramidite Approach), 17 (Seliger H. Scale-Up of Oligonucleotide Synthesis: Solution Phase) and 18 (Sinha N. Large-Scale Oligonucleotide Synthesis Using Solid -Phase Approach). S Agrawal, ed. Humana Press, 1993, and references cited therein.

Protocols for Oligonucleotide Conjugates. Chapters 9 (Warren WJ, Vella G. Analysis and Purification of Synthetic Oligonucleotides by HPLC), 10 (Black DM, Gilham PT. Sequence Analysis of Oligodeoxyribonucleotides), 11 (Andrus A. Gel-Capillary Electrophoresis Analysis of Oligonucleotides), 12 (Jaroszewski JW, Roy S, Cohen JS. NMR Studies of Oligonucleotides), 13 (McClure TD, Schram KH. Mass Spectrometry of Nucleotides and Oligonucleotides). S Agrawal, ed Humana Press, 1994, and references cited therein.

Rao KVB, Chiu Y-Y, Chen CW, Blumenstein JJ. Regulatory concerns for the chemistry, manufacturing, and controls of oligonucleotide therapeutics for use in clinical studies. *Antisense Res Dev* 1993;3:405-10.

Ravikumar VT, Krotz AH, Cole DL. Efficient synthesis of deoxyribonucleotide phosphorothioates by the use of DMT cation scavenger. *Tetrahedron Lett* 1995;36:6587-90.

Reddy MP, Farooqui F, Hanna NB. Methylamine deprotection provides increased yield of oligoribonucleotides. *Tetrahedron Lett* 1995;36:8929-32.

Reese CB, Song Q. Avoidance of sulfur loss during ammonia treatment of oligonucleotide phosphorothioates. *Nucl Acid Res* 1997;25:2943-44.

Sanghvi YS, Cook PD. "Carbohydrates: Sythetic Methods and Applications in Antisense Therapeutics: An Overview" In *Carbohydrate Modifications in Antisense Research*. ACS Symposium Series 580. YS Sanghvi and PD Cook, eds. ACS Publications, 1994.1-22.

Sanghvi YS. "DNA with altered backbones in antisense Applications" In *Comprehensive Natural Product Chemistry*, Barton DHR, Nakanishi K. (editor-in-chief) In *DNA and Aspects of Molecular Biology*, ET Kool, ed. Elsevier Science Ltd., 1998, In press.

Schuette JM, Pieles U, Maleknia SD, Srivatsa GS, Cole DL, Moser HE, Afeyan NB. Sequence analysis of phosphorothioate oligonucleotides via matrix-assisted laser desorption ionization time-of-flight mass spectroscopy. *J Pharm Biomed Analys* 1995;13:1195-203.

Srivatsa GS, Batt M, Schuette J, Carlson R, Fitchett J, Lee C, Cole DL. Quantitative capillary gel electrophoresis assay of phosphorothioate oligonucleotide in pharmaceutical formulations. *J Chromatogr A* 1994;680:469-77.

Srivatsa GS, Klopchin P, Batt M, Feldman M, Carlson RH, Cole DL. Selectivity of anion exchange chromatography and capillary gel electrophoresis for the analysis of phosphorothioate oligonucleotides. *J Pharm Bio Med Analys* 1997;16:619-30

Srivatsa GS. Analysis and control of synthetic oligonucleotides: past, present, and future. Large-scale oligonucleotide synthesis. *IBC meeting*, 1997 October 28-29, San Diego, California.

Vargeese C, Carter J, Yegge J, Krivjansky S, Settle A, Kropp E, Peterson K, Pieken W. Efficient activation of nucleoside phosphoramidites with 4,5-dicyanoimidazole during oligonucleotide synthesis. *Nucleic Acids Res* 1998;26:1046-50.

Warren WJ, Vella G. Analysis and purification of synthetic oligonucleotides by HPLC. in *Protocols for Oligonucleotide Conjugates, 1994*, 233-264.

Wyrzykiewicz TK, Cole DL. Sequencing of oligonucleotide phosphorothioates based on solid-supported desulfurization. *Nucl Acid Res* 1994;22:2667-69.

Zhang Z, Tang JY. A novel phosphitylating reagent for in situ generation of deoxyribonucleoside phosphoramidites. *Tetrahedron Lett* 1996;37:331-34.

1.6 NOTES

†   Lars Holmberg is at Amersham Pharmacia Biotech, Milwaukee, WI 53202.
*   Author to whom the correspondende should be sent
1. Amersham Pharmacia Biotech, Piscataway, NJ (1-800-526-3593).
2. Millipore Corp., Bedford, MA (1-800-645-5476).
3. Hybridon, Milford, MA (1-888-DNA-HYBN).
4. ABI-Perkin Elmer, Foster City, CA (1-800-874-9868).
5. PerSeptive Biosystems, Framingham, MA (1-800-899-5858).
6. Advanec Zoyo Kaisho Ltd., Japan.

7. Biotage, Division of Dyax Corp., Charlottesville, VA (1-800-446-4752).
8. Waters Corp., Milford, MA (1-800-252-4752).
9. Carr Separations, Inc. Franklin, MA (508-553-2400).
10. Leybold Heraus, Germany (AMSCO).
11. Finnigan MATT, San Jose, CA.
12. Beckman Instruments, Fullerton, CA (see reference 32 for details).
13. Voyager: PerSeptive Biosystems, Cambridge, MA or LDI 1700: Linear Scientific Inc., Reno, NV (see reference 30 for details).
14. Unity 400: Varian Associates, Palo Alto, CA.
15. Gilford Response II, Oberlin, Ohio.
16. Further details on this aspect can be directly requested from APB (note 1).
17. This also enables calculations of % coupling efficiency of the previous cycle and thus overall yield of AS ODN synthesis.
18. This reagent is named after its inventor, S. L. Beaucage.
19. Incomplete sulfurization of **4** will result in subsequent cleavage of phosphite triester bond and formation of a truncated sequence.
20. Z. Cheruvallath *et.al.* manuscript in preparation. Use of **9** was first reported by J. H. van Boom *et.al. Tetrahedron Lett.* 1989; 30:6757-6760.
21. If the mixing of the two capping solutions is not efficient as they reach the synthesis support, there is a risk of detritylation caused by the acidity of the Ac_2O solution alone.
22. BioSepra, Marlborough, MA (1-800-752-5277).
23. Conditions for purification of 5320: DMT-off ISIS 5320 crude was separated on POROS HQ 50 media using IE buffers mentioned in Section 2.2.

2 How to Choose Optimal Antisense Targets in an mRNA

Bernhard Schu and Heinrich Brinkmeier

Department of General Physiology
University of Ulm
Ulm, Germany

2.1 INTRODUCTION

One of the major challenges of an antisense experiment is the identification of sequences within the target RNA, which are accessible by an antisense oligonucleotide. Three-dimensional folding of the RNA results in both inaccessible segments in the inner part of the RNA molecule and double stranded regions. The folding of RNAs may prevent many oligonucleotides from reaching their complementary regions (Branch, 1998). While it is impossible to predict accessible target sites for antisense oligonucleotides from computer-aided models of RNA-structure (Fig.2), combinatorial screening is a promising approach to search for such sequences. The identification of the most active antisense oligonucleotide however remains empirical and has to be performed in translation arrest assays and in cell culture. Sequences with well-known non-antisense effects have to be avoided. Finally, stringent controls are required to prove an antisense-mediated mechanism of action.
(Stein, 1995; Branch, 1996)

For both the screening for the target and the validation of the antisense effect the use of in vitro systems are highly recommendable.

In this chapter we present a systematic strategy to obtain effective antisense oligonucleotides for a given target RNA (see Fig. 1). Furthermore, we discuss reliable control experiments as well as additional methods to optimize antisense oligonucleotides.

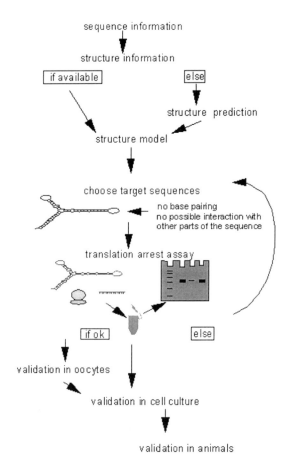

Figure 1: Schematic representation of target screening

2.2 COMBINATORIAL SCREENING FOR TARGET SITES

There are recently developed techniques to establish libraries containing several thousands of different oligonucleotides. In some cases these oligonucleotides are fixed on a surface, in others they are in solution. Such libraries can be used to search for accessible antisense target sites.

Radioactively labeled and natively folded RNA can be hybridized against such a library. For a particular RNA molecule the best binding oligonucleotides can be identified. By this strategy called combinatorial screening one has direct information about the accessibility of oligonucleotides to an RNA target site.

A recent paper of Lima et al. describes a strategy in solution (Lima et al., 1997). The authors use radioactive endlabeled mRNA, a random library of all 10mers and RNase H. The oligonucleotides hybridize against all accessible parts of the RNA and RNase H cleaves the RNA at each of these sites. Subsequent gel electrophoresis is carried out to detect the cleavage sites. All of them are potent sites for antisense oligonucleotide action.

This procedure certainly needs a lot of experimental skill by means of RNA endlabeling and producing reproducible results with the RNase H. It is recommendable for groups which plan to work with the antisense technique on a lot of different RNAs.

The group of Southern at the University of Oxford U.K. has introduced a method to construct oligonucleotide libraries on glass surfaces (Milner et al., 1997). They propose a hybridization strategy to find accessible antisense target sites. Therefore, they produced a library of 1938 on a defined place on the solid surface. Folded RNA was radiolabeled by introducing $[\alpha^{32}P]rUTP$ and hybridized against the library. All oligonucleotides that are bound by the RNA are potential candidates for the antisense experiment.

Bader et al. have developed a method to create a one-dimensional library of overlapping oligonucleotides of a particular sequence immobilized on polypropylene (Bader et al., 1997). The library starts at any interesting point of the sequence with the first 15-mer (being complementary to the first 15 residues from the starting point) and "walks" along the sequence by proceeding one nucleotide and building the next 15-mer. This can be repeated several times resulting in a library of some complementary 15-mers. By hybridization of labeled RNA against these oligonucleotides a part of its sequence should be scanned for accessibility by antisense oligonucleotides. Although in this system the hybridization behavior is shown only with deoxyoligonucleotides it is a promising tool for the search of an optimal antisense oligonucleotide against one particular point mutation.

The field is developing rapidly, and DNA microarrays will soon be available commercially. For further reading we refer to (Schena, 1999).

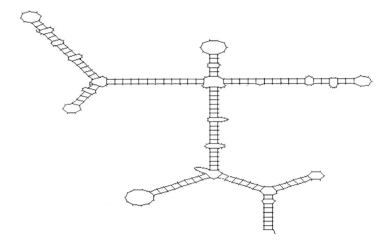

Figure 2: Scheme of an RNA structure displayed as "squiggles" The energy minimum of a particular RNA folding was calculated using the algorithm of Zucker and displayed as stems and loops. RNA modeling has not been proven useful for the design of antisense oligonucleotides.

2.3 SCREENING OF ANTISENSE TARGET SITES BY IN VITRO EXPERIMENTS

The selection of oligonucleotides merely based on the prediction of loop segments is often not sufficient to obtain oligonucleotides of optimum effectiveness. Thus, it is strongly recommendable to verify the effects of preselected oligonucleotides by experimental means. Several in vitro methods are available to test the accessibility of an oligonucleotide to its target site and to check for its inhibitory effect on protein translation.

RNA synthesis

The first step for the in vitro experiments with the chosen oligonucleotides is the preparation of the respective RNA by in vitro transcription. The handling of RNA is slightly crucial because of the nuclease sensitivity of RNA. Care must be taken to avoid contamination with RNases (Sambrook et al., 1987). Use only RNase-free buffers, water, supplies and wear gloves during experimental procedures. One crucial step in the protocols listed below to avoid the decay of RNA is phenol extraction of protein. This purification step results in protein free solutions and consequently in RNase-free environment.

If RNA shall be stored over a longer period of time precipitate and keep it under 70% ethanol at –70 °C.

Preparing DNA for in vitro transcription

A broad assortment of vectors is available which contain a promoter site for SP6, T3, T7 RNA polymerase, respectively. Each of these plasmids can be used for in vitro transcription. Subclone your cDNA into such a vector and linearize it with a suitable restriction enzyme as a template for the RNA polymerase.

Alternatively, DNA templates for in vitro transcription can be prepared quite conveniently by PCR. Two PCR-primers should be designed that flank the coding region of the gene of interest. In most cases it is not necessary for a sufficient translation to include the 5' and 3' UTRs into the PCR product and the RNA respectively.

A suitable promoter has to be introduced into the amplification product with the upstream primer, which has to contain one of the following sequences additionally to the complementary part of the cDNA (Logel et al., 1992).

SP6 promoter	AAT TAG GTG ACA CTA TAG AAT AG
T3 promoter	AAT TAA CCC TCA CTA AAG GGA AG
T7 promoter	TAA TAC GAC TCA CTA TAG GGA GA

Experimental procedure

• *Mix:*

5µl	10x Taq buffer	(containing 15 mM Mg^{2+})
1µl	DNA	(containing 3×10^5 molecules)
1µl	primers	(final concentration 20 pmol each)
2 U	Taq polymerase	

water to a final volume of 50 µl

- Incubate under the following cycling conditions (optimized for PCR machine Model PTC-150 MJ Research).

step	temperature	duration
1	95 °C	40 s
2	92 °C	5 s
3	50 °C	45 s
4	72 °C	60 s
5	go to step 2, 4 more times	
6	92 °C	5 s
7	60 °C	45 s
8	72 °C	60 s
9	go to step 6, 24 more times	
10	72 °C	60 s
11	4 °C	∞

- Purify by phenol extraction and analyze the products by gelelectrophoresis.

In vitro transcription (Protocols and Application Guide, 1996)

Use the linearized and purified DNA as template for a run off transcription. Note that the stability of the ribonucleotides is crucial. Therefore aliquot them and store at -70°C.

For suitable translation of the RNA it is mostly essential to perform the transcription in presence of m7G(5')ppp(5')G, which is used as a primer by the RNA polymerase resulting in a capped RNA.

Experimental procedure *

• Mix at room temperature:

---

4 μl	transcription buffer	(5x)
2 μl	DTT	(100 mM)
20 U	RNasin	
4 μl	ATP, CTP, UTP	final concentration 5 mM each
2 μl	GTP	10 mM
5 μl	m7G(5')ppp(5')G	(5 mM)
1 μl	linearized template DNA	(1 μg/μl)
40 U	RNA polymerase	(SP6, T3, T7 according to the promoter)
	water to a final volume of 20 μl	

---

* refers to a kit provided by Promega Corporation Madison, WI, USA

• Incubate 60 to 120 min at 37°C
• Stop the reaction by adding 2U DNase (RNase-free) and incubate15 min at 37°C
• Purify by phenol extraction.

Testing for RNA integrity:

Incubate the reaction in presence of [α^{32}P]rUTP followed by denaturing PAGE (4-6%, 7M urea). Alternatively, non radioactive RNAs can be separated with TAE agarose gels containing 5 mg/l iodoacetic acid and stained with ethidiumbromid.
    This control should be performed in each case to be sure that an intact RNA is used for the following experimental steps.

Note:
If shorter transcripts appear optimize the reaction. In such cases lowering the temperature or lowering the concentration of the DNA template often lead to success. Be sure that your template DNA is quantitatively linearized, otherwise very long RNAs can be built resulting in a rapid decrease of the concentration of the nucleotides and consequently a very inhomogeneous RNA population.
    An elongated incubation time (up to 4 hours) with the addition of RNA polymerase after 2 hours may improve the RNA yield.

Low yield in RNA may result from low concentrations of DNA and high concentrations of salt (e.g. NaCl). DNA may precipitate in presence of spermine (contained in the transcription buffer). Be sure to mix the reaction at room temperature. Salt may result from former DNA precipitation. Do not precipitate at -20 °C. Precipitation at room temperature or on ice will work fine and results in lower salt contamination. Wash precipitated DNA extensively with 70% ethanol (room temperature).

Translation arrest assay

In vitro translation in cell free systems (Protocols and Application Guide, 1996)

There are two different cell free systems at hand that can be used to translate RNA into proteins in the test tube: Rabbit reticulocyte lysate and wheat germ extract.

Both systems contain all cellular components necessary for a biosynthesis of proteins except mRNAs, which have been removed by micrococcal nuclease, e.g. to reduce background translation. Thus, the addition of one single RNA results in the production of one single protein.

RNase H, which certainly plays a major role in the inhibitory effect of antisense oligonucleotides (see below), is present in either cell-free system, but in higher concentrations in wheat germ extract. In reticulocyte lysate the RNase H concentration may be variable and sometimes even close to zero depending on the batch obtained. For that reason antisense oligonucleotides are active at much lower concentrations in wheat germ extract. Nevertheless, rabbit reticulocyte lysate is a blameless system to test them in vitro if one has in mind the higher concentration of oligonucleotides needed. It is even possible to enhance the inhibitory effect in the rabbit system by adding RNase H to the reaction mixture (Cazenave et al., 1989).

In some cases one particular protein is translated poorly in one of both systems so that one has the chance to improve the results by switching to the other.

Usually radioactive [^{35}S]methionine is added to the reaction mixture to label the newly synthesized protein, which can be separated by gelelectrophoresis and detected by autoradiography. In such a system the action of antisense molecules can be studied very nicely because cross reactions are almost excluded in this first step. Nevertheless, also in this stage control reactions (see below) must be performed very accurately to confirm a real antisense effect.

Alternatively, non radioactive detection systems are available containing a biotinylated aminoacyl-tRNA (for lysine) to introduce biotin as a label into the protein.

Note that some postranslationally modified proteins may lack methionine and so [^{35}S]cysteine or another modified amino acid must be used to label the product. In some cases, i.e. if proteins are membrane bound, microsomal membranes must be added to get a satisfactory yield.

In vitro translation systems are very sensitive to oxidation. Thus, they should be aliquoted immediately after purchasing and stored at −70 °C. Repeated thawing and freezing should be avoided. Different preparations of the reticulocyte lysate or wheat germ extract may have different translation activity, so use one lot of the in vitro translation kit for a single antisense project.

*Experimental procedure**

• Mix:

16 µl	rabbit recticulocyte lysate or wheat germ extract	
20 U	RNasin	
0,5 µl	amino acid mixture, minus methionine	1 mM
40 µCi	[^{35}S]methionine	
5 µl	oligonucleotide solution	(final concentration 1 to 10 µM)
1 µl	capped RNA transcript	(0.1 –1 µg, depending on RNA length, optimize empirically)
100 U	RNase H	optional
	water to a final volume of 25 µl	.

* refers to a kit provided by Promega Coorporation Madison, WI, USA

• Incubate for 120 min at 33 °C.
• Stop the reaction on ice
• Add an equal volume of 2x sample buffer (125 mM Tris-HCl pH 6.8, 6% SDS, 0.2 g bromphenolblue, 20% glycerol, 5% mercaptoethanol) and heat 3 min to 95 °C to denature the proteins. If proteins will precipitate, caused by the 95 °C step, just add the 2x sample buffer to the reaction mixture at room temperature.
• Place on ice.
• An aliquot of 5-10 µl can be applied directly to a SDS polyacrylamid gel (12-15% acrylamid).

Notes:

We use exclusively rabbit reticulocyte lysate for our experiments. Good oligonucleotides show complete inhibition at 10 µM (no protein band detectable), a strong inhibition at 1 µM (a weak band is detected), and even at 0.32 µM the inhibition can be observed. Have in mind that in wheat germ extract the antisense effect is stronger due to the higher concentrations of RNase H.

The relation between RNA concentration and the concentration of antisense oligonucleotides plays an essential role in this translation arrest assay. The data given above are optimized for an RNA of about 500 bases and a 15-mer oligonucleotide.

To avoid oxidation of ^{35}S-labeled amino acids store them in aliquots with buffers containing DTT or β-mercaptoethanol at -70°C.

Do not add calcium to the reaction mixture. Calcium could reactivate nuclease activity resulting in the destruction of the RNA.

The addition of spermine to a final volume of (0.1-0.5 mM) may stimulate the translation and can help optimizing the reaction (Snyder et al., 1991).

If a strong background appears, incubation with RNase A (0.2 mg/ml final concentration) for 5 min at 30°C may reduce it by digesting the tRNAs.

The translation efficiency may also be optimized by adding phosphocreatine kinase and phosphocreatine as an energy delivery system, higher concentrations of tRNA for the translation of very long transcripts, and, finally, magnesium or potassium acetate depending of the particular mRNA.

Xenopus oocytes

In some cases antisense studies in Xenopus oocytes may be very helpful for validation of the antisense effect rather then screening for target sites. With this little bioreactors one has a complete living environment, including RNase H, (Shutteworth et al., 1988) activity which is quite easy to handle. RNA can be injected alone or together with antisense oligonucleotides directly into the oocytes and will be translated into protein. Physiologists use this system routiniously to express and study ion channels in a heterologous system (Zühlke et al., 1995).

The advantage of oocytes for validation of antisense oligonucleotides compared to other cells beside their size is that, by direct injection, there is no doubt that the oligonucleotide reaches the cytoplasm and is not retarded by the plasma membrane or captured in lysosomes.

Experimental procedures

For details in preparation of oocytes see Zühlke et al (Zühlke et al., 1995).

Inject varying amounts of RNA with or without antisense oligonucleotides into the oocytes. After 3 days the protein can be detected. The amounts of oligonucleotides required varies with the length of the RNA.

We used this approach with antisense oligonucleotides directed against the mRNA of the sodium channel of human skeletal muscles (hSKM1) e.g. In that setup we injected 1.25 µg (in 10 nl). When antisense oligonucleotides were coinjected they were mixed 1:1 (v/v) with the RNA in a test-tube (RNA conc.: 0.25 µg/µl, oligo conc.: 200µM in H_2O), 10 nl were injected. The size of RNA in our case was about 6000 basepairs, the oligos were 15 mers (Brinkmeier et al., 1997).

2.4 CONTROL EXPERIMENTS TO VALIDATE ANTISENSE EFFECTS

An antisense effect has to be controlled intensively because several non-antisense effects may occur. Different control experiments are helpful in different approaches. In general, it is preferable to perform more than one kind of control. All of these controls can be performed in both translation arrest or RNase H assays (see below).

- Cross reactions

One outstanding experimental control is the addition of an RNA coding for a closely related isoform of the protein studied. If the antisense oligonucleotide distinguishes between both by means of suppressing the translation of the target without interfering with the other RNA you have a nice proof for the sequence specificity. Because of the relationship other non-antisense effects can by widely excluded.

- Mismatched oligonucleotides

Oligonucleotides with one, two, or three mismatches serve as very sensitive control molecules. The influence of these mismatches on the target specificity of antisense oligonucleotides can be depicted directly in translation arrest assays. One and two unpaired nucleotides should diminish the inhibitory potential of the oligonucleotide broadly three mismatches should abolish it.

- Mismatched target sites

Target sites can be altered by in vitro mutagenesis. Such a strategy allows to simulate the behavior of antisense oligonucleotides on point mutations versus wildtype.

- Inverted oligonucleotides

In this control reaction only the polarity of the oligonucleotide is changed.

- Sense oligonucleotides

Sense oligonucleotides can be used in control experiments to proof a real antisense effect. This kind of control reaction bears the risk that sense oligonucleotides may bind to a second target site within the RNA which is complementary to the actual point of antisense attack. In this case both oligonucleotides may inhibit the translation pretending a non- or less active antisense oligonucleotide.

- Scrambled oligonucleotides

The sequence of the oligonucleotide is reorganized in a random way maintaining the base composition. available.

- Concentration dependence

In a dose response curve a clear dependence of the antisense effect on the oligonucleotide concentration should be visible.

Inverted or sense oligonucleotides are not recommended. Similar control reactions can be used to verify the antisense effect in cells or in vivo.

2.5 ALTERNATIVE METHODS FOR THE SCREENING OF TARGET SITES

RNase H assay to screen for accessible target sites on the RNA

One major mechanism even if not the exclusive one of antisense inhibition of translation is the degradation of RNA in the mRNA/antisense oligonucleotide duplex by RNase H. The RNA/DNA heteroduplex is substrate of that enzyme which cleaves within the RNA part. This results in a truncated mRNA which does not longer code for an intact protein and moreover is fast degraded by cellular Rnases (Boiziau et al., 1991).

Thus, it is obvious to use RNase H in vitro to test the accessibility both of the oligonucleotide to RNA and of the enzyme to the duplex. RNase H can be purchased commercially and incubated in an RNase H buffer together with RNA and oligonucleotide. For that study it is advantageous to have radioactively labeled RNA (incubate the transcription reaction in presence of $[\alpha^{32}P]rUTP$) so that the cleavage product can be analyzed by PAGE followed by autoradiography. The minimum length of the DNA part in the RNA/DNA duplex must be 5 residues. This is especially important if modified oligos shall be used. Only unmodified or phosphorothioate oligos exhibit RNase H stimulation, other modifications or RNA oligos don't. Thus, if RNase H activity is desired a minimal stretch of unmodified or phosphorothioate nucleotides of 5 bases in length is necessary.

Experimental procedure (Donis Keller, 1979)
- Mix on ice

2,5 µl	10x RNase H buffer	(200 mM HEPES pH 8.0, 10 mM DTT, 100 mM MgCl$_2$, 500 mM KCl)
1 µg	Native RNA	radiolabeled
	Antisense oligonucleotide	10x molar excess
1 U	RNase H	
	Water to a final volume of 25 µl	

- Incubate at 37°C for 60 min. Stop the reaction by adding 1µl EDTA (500 mM pH 8.0) and analyze products by PAGE.

2.6 ACKNOWLEDGMENTS

The authors would like to thank the IZKF (interdisciplinary center of medical research) Ulm and the University of Ulm for financial support.

2.7 REFERENCES

Bader R, Brugger H, Hinz M, Rembe C, Hofer EP, Seliger H. A rapid method for the preparation of a one dimensional sequence-overlapping oligonucleotide library. *Nucleot Nucleos* 1997;16:835-42

Branch AD. A hitchhiker's guide to antisense and nonantisense biochemical pathways. *Hepatology* 1996;24:1517-29

Branch AD. A good antisense molecule is hard to find. *Trends Biochem Sci* 1998;23:45-50

Brinkmeier H, Schu B, Seliger H, Kurz LL, Buchholz C, Rudel R. Antisense oligonucleotides discriminating between two muscular Na+ channel isoforms. *Biochem Biophys Res Commun* 1997;234:235-41

Boiziau C, Kurfurst R, Cazenave C, Roig V, Thuong NT, Toulme JJ. Inhibition of translation initiation by antisense oligonucleotides via an RNase-H independent mechanism. *Nucleic Acids Res* 1991;19:1113-19

Cazenave C, Stein CA, Loreau N, Thuong NT, Neckers LM, Subasinghe C, Helene C, Cohen JS, Toulme JJ. Comparative inhibition of rabbit globin mRNA translation by modified antisense oligodeoxynucleotides. *Nucleic Acids Res* 1989;17:4255-73

Donis Keller H. Site specific enzymatic cleavage of RNA. *Nucleic Acids Res* 1979;7:179-92

Lima WF, Brown DV, Fox M, Hanecak R, Bruice TW. Combinatorial screening and rational optimization for hybridization to folded hepatitis C virus RNA of oligonucleotides with biological antisense activity. *J Biol Chem* 1997;272:626-38

Logel J, Dill D, Leonard S. Synthesis of cRNA probes from PCR-generated DNA. *Biotechniques* 1992;13:604-10

Milner N, Mir KU, Southern EM. Selecting effective antisense reagents on combinatorial oligonucleotide arrays. *Nat Biotechnol* 1997;15:537-41

Protocols and Application Guide. 1996. Promega Corporation

Sambrook J, Fritsch EF, Maniatis T. *Molecular cloning: a laboratory manual.* 1987. Cold Spring Habor Laboratory Press, New York

Shuttleworth J, Colman A. Antisense oligonucleotide-directed cleavage of mRNA in xenopus oocytes and eggs. *EMBO J* 1988;7:427-34

Schena M (ed). DNA Microarrays: A Practical Approach. IRL press, 1999 (in press).

Snyder RD, Edwards ML. Effects of polyamine analogs on the extent and fidelity of in vitro polypeptide synthesis. *Biochem Biophys Res Commun* 1991;176:1383-92

Stein CA. Does antisense exist? *Nat Med* 1995;1:1119-21

Zühlke, RD, Zhang HJ, Joho RH. Xenopus oocytes: a system for expression cloning and structure-function studies of ion channels and receptors. *Method Neurosci* 1995;25:67-89

Part II:
Antisense Application
in vitro

3 How to Characterize and Improve Oligonucleotide Uptake into Leukocytes

Martin Bidlingmaier[1], Anne Krug[2]
and Gunther Hartmann[2]

[1]Divisions of Neuroendocrinology and
of [2]Clinical Pharmacology
Medizinische Klinik, Klinikum Innenstadt
of the Ludwig Maximilians University
Munich, Germany

3.1 INTRODUCTION

White blood cells, which consist of granulocytes, monocytes/macrophages and different subpopulations of lymphocytes, are important target cells for antisense strategies, especially in the field of immunology and infection. Adhesion molecules and cytokines, which are produced by leukocytes, regulate inflammatory responses and are overexpressed in inflammation and infection. Antisense oligonucleotides, which specifically inhibit the production of these essential molecules, are promising drugs in the treatment of chronic inflammatory diseases, such as ulcerative colitis, M. Crohn or rheumatoid arthritis. Inhibition of ICAM-1 (intercellular-adhesion-molecule-1) by antisense oligonucleotides reverses experimental colitis in mice and leads to reduction in inflammation and clinical benefit in patients with Crohn's disease, as recent results of a phase II clinical study show (Bennett et al., 1997; Bradbury, 1997). Specific suppression of monocyte/macrophage-derived cytokines, such as tumor necrosis factor (TNF)-α, Interleukin (IL)-1β, IL-6 and IL-10, by antisense oligonucleotides has been demonstrated in cell culture experiments (Hartmann et al., 1996; Keller and Ershler, 1995; Peng et al., 1995; Yahata et al., 1996). Moreover, T-helper lymphocytes and monocytes/macrophages are target cells for antisense strategies against HIV-infection (Lund et al., 1995; Pirruccello et al., 1994).

Hematopoetic tumor cells of the myeloid and lymphoid lineage are targeted by antisense sequences for specific inhibition of oncogene expression. Antisense oligonucleotides directed against the anti-apoptosis protooncogene bcl-2 have been shown to suppress bcl-2 protein expression and leukemic cell growth (Reed et al., 1990). Early results of a phase I clinical study in patients with non-Hodgkin lymphoma suggest antitumor activity of an anti-bcl-2 oligonucleotide (Webb et al., 1997).

These data show that human hematological cells are targets for antisense strategies in the treatment of inflammatory, infectious and neoplastic diseases. Moreover, circulating blood cells are unintentionally exposed to systemically administered oligonucleotides. It is therefore important to investigate to what extent and by which mechanism oligonucleotides are incorporated into human blood cells.

Oligonucleotide uptake into cells: basic mechanisms and approaches for improving oligonucleotide delivery

As antisense oligonucleotides are large polyanionic molecules, poor cellular uptake and access to their site of action remain major obstacles for the application of antisense oligonucleotides. Phosphodiester- as well as phosphorothioate-oligonucleotides are spontaneously incorporated into cells either by fluid phase endocytosis (for higher concentrations) or receptor-mediated endocytosis (for lower concentrations (Beltinger et al., 1995). This process is saturable and dependent on incubation time, temperature, oligonucleotide concentration and length, but independent of sequence (Loke et al., 1989; Temsamani et al., 1994). Backbone modifications in the sugar or phosphate moiety, as for example phophorothioate-modification, influence oligonucleotide uptake characteristics. Spontaneous oligonucleotide incorporation is also strongly dependent on cell type as well as activation and differentiation state of the cells. For example, oligonucleotide uptake is lower and more heterogeneous in primary cells than in permanent cell lines. Several groups found that oligonucleotides are localized in endosomal structures following spontaneous incorporation into the cells (Bennett et al., 1992; Zhao et al., 1993; Zhao et al., 1994). Studies performed in certain cell lines, in which specific suppression of the target protein by exogeneously applied antisense oligonucleotides could not be demonstrated, suggest that the antisense molecules do not reach their proposed site of action in effective concentrations at least in these cell types (Hartmann et al., 1996; Matteucci and Wagner, 1996).

Various methods for improving oligonucleotide uptake and intracellular distribution have been developed. Microinjection of oligonucleotides into the cytoplasm of cells is followed by nuclear accumulation and allows antisense mediated suppression of the target protein (Fisher et al., 1993; Wagner et al., 1993). Similarly, cationic lipids form complexes with DNA-molecules and thus facilitate transport of oligonucleotides across the cell membrane which is followed by release into the cytoplasm and rapid accumulation in the nucleus (Bennett et al., 1992). Several *in vitro* studies demonstrate that the use of cationic lipids for oligonucleotide delivery is essential for achieving a specific antisense effect (Bennett et al., 1992; Hartmann et al., 1996; Vaughn et al., 1995; Wagner et al., 1993). Other methods such as liposomal encapsulation or electroporation have also shown to be effective for oligonucleotide delivery into target cells (Chavany et al., 1995; Wang et al., 1995).

Methods for investigation of cellular oligonucleotide uptake

Different methods can be used to the quantify cellular oligonucleotide-uptake. One possibility is the use of radio-labeled oligonucleotides followed by measurement of cell associated radioactivity or analysis of extracted oligonucleotides by gel-electrophoresis. Oligonucleotides labeled at each internucleotide linkage by incorporation of ^{35}S (in completely phosphorothioate modified sequences) or ^{32}P-end-labeled oligonucleotides have been used for this purpose (Iversen et al., 1992; Krieg et al., 1991; Temsamani et al., 1994). Another method is the detection of unlabeled oligonucleotides extracted from cells by blotting and hybridization with a complementary radio-labeled oligonucleotide probe (Temsamani et al., 1993; Zhao et al., 1996). This method guarantees that cellular oligonucleotide association is not altered by the label. Both methods however do not allow differentiation between oligonucleotides associated with dead or with living cells.

Flow cytometry as a method for quantification of cell-associated oligo-nucleotides that are fluorescence-labeled at one end has been applied in several studies (Iversen et al., 1992; Krieg et al., 1991; Marti et al., 1992; Pirruccello et al., 1994; Zhao et al., 1993; Zhao et al., 1996; Zhao et al., 1994). This method provides the possibility to distinguish cell-subpopulations expressing specific antigens and to exclude dead cells from analysis. It has been argued that detected fluorescence caused by labeled oligonucleotide degradation products or free fluorescence-labels might be mistaken for intact oligonucleotides. Zhao et al. performed uptake studies with fluorescence-labeled and unlabeled oligonucleotides comparing detection by flow cytometry with detection by extraction and slot-blot. Corresponding results regarding the time course of oligonucleotide uptake over 24 hours were obtained with both methods (Zhao et al., 1996). This confirms the reliability of the flow cytometric method.

Several approaches have been chosen to characterize the localization of cell-associated oligonucleotides. Cell fractionation studies have been performed to investigate the distribution of radio-labeled oligonucleotides in different cellular compartments (Beltinger et al., 1995; Iversen et al., 1992; Temsamani et al., 1994). Alternatively, fluorescence microscopy can be used to determine localization of fluorescence-labeled oligonucleotides. In contrast to cell fractionation studies, fluorescence-microscopy can be performed with suspensions of vital cells parallel to flow-cytometric quantification of cellular oligonucleotide-association (Albrecht et al., 1996; Hartmann et al., 1998b). In addition to conventional microscopy, confocal laser scanning microscopy allows exact differentiation of intracellular and surface-bound fluorescence, as well as localization of oligonucleotides in subcellular compartments. Similarly to flow-cytometric analysis, this methods allows parallel staining of cell subpopulations by specific fluorescence-labeled antibodies (Hartmann et al., 1998b).

The combination of flow cytometry and fluorescence microscopy appears to be a suitable method for the investigation of cellular oligonucleotide uptake and intracellular distribution. The influence of different incubation conditions and uptake enhancing methods on oligonucleotide incorporation can thus be evaluated. Detailed protocols for the examination of oligonucleotide-uptake and intracellular localization in human peripheral blood cells using these methods are described in the following.

3.2 EXPERIMENTAL APPROACHES

Preparation of cells

There are several approaches for the preparation of human leukocytes from peripheral blood for the investigation of cellular oligonucleotide-uptake. Mononuclear cells, consisting of lymphocytes and monocytes, can be isolated from human peripheral blood by density gradient centrifugation and then be re-suspended in cell culture medium containing FITC-conjugated oligonucleotides. This approach allows for identical incubation conditions independent of variable plasma components of individual blood donors. It cannot reflect the *in vivo* situation of circulating leukocytes, however, since isolated mononuclear cells are deprived of granulocytes, erythrocytes, thrombocytes and plasma, in other words, their natural environment in peripheral blood. Another approach, which comes closest to the *in vivo* situation is the incubation of anticoagulated whole blood samples with FITC-oligonucleotides. Both methods are described in detail below. Preparation of cells and subsequent incubation with oligonucleotides should be performed under sterile conditions. To avoid activation of monocytes by exposure to endotoxin use sterile tubes, work under laminar flow and use fresh endotoxin-tested reagents. Adhesion of monocytes, which also leads to activation, can be minimized by using polypropylene tubes during cell preparation and subsequent steps.

Peripheral blood mononuclear cells

Ten milliliters of blood roughly equal a yield of 10×10^6 mononuclear cells consisting of approximately 20 % monocytes and 80 % lymphocytes. Anticoagulate blood with 40 I.U./ml heparin (Braun, Melsungen, Germany). Fill 15 ml of Ficoll-

Hypaque (Biochrom, Berlin, Germany) into 50 ml Leucosep® tubes (Greiner, Frickenhausen, Germany), which contain a horizontal porous filter disc to facilitate layering of blood, and centrifuge (1000 g, 20°C, 2 min). Add a 15 ml layer of blood onto the filter disc and dilute with 20 ml 0,9 % NaCl (room temperature). Centrifuge (1000 g, break off, 20°C, 20 min), aspirate the mononuclear cell layer and dilute with 50 ml 0,9 % NaCl in 50 ml polypropylene tube (Becton Dickinson, Lincoln Park, USA). Pellet cells (500 g, 4 °C, 10 min) and resuspend in 0,9 % NaCl (4 °C) twice. Then pellet and resuspend cells in prewarmed CO_2-equilibrated RPMI 1640 culture medium (Biochrom) supplemented with 2 mM L-glutamine and 10 mM HEPES (Sigma, Munich, Germany). Use serum-free medium for incubation of PBMC with oligonucleotides and lipofectin, as both agents have been reported to interact with serum proteins (Felgner et al., 1993; Srinivasan et al., 1995). Determine cell viability by trypan blue exclusion and adjust cell density to 10×10^6 cells/ml.

Whole blood leukocytes

Anticoagulation is the critical step when using native whole blood. We found that EDTA and heparin both decrease oligonucleotide uptake into blood cells in a concentration dependent manner (see fig. 1). The inhibiting effect of EDTA on oligonucleotide uptake is obviously caused by complexation of Ca^{2+} ions to EDTA as it could be reversed by the addition of calcium, but not magnesium (Hartmann et al., 1998a). (1) heparin and oligonucleotides are polyanionic molecules. (2) Leukocytes bind to polyanionic molecules of the extracellular matrix (i.e. hyaluronic acid). (3) Thus, surface binding and subsequent uptake of oligonucleotides might be competed by heparin.

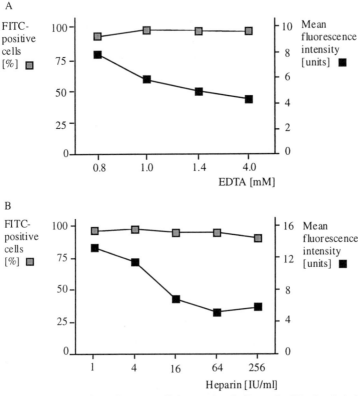

Figure 1: Anticoagulants decrease cellular uptake of oligonucleotides in whole blood: Whole blood samples were incubated for 2 h with FITC-labeled oligonucleotide (500 nM) in the presence of increasing concentrations of EDTA (A) or heparin (B). Oligonucleotide uptake into monocytes is shown. Monocytes were identified by staining with PE-labeled antibody against CD14. Oligonucleotide incorporation was quantified by the percentage of FITC-positive cells (gray symbols, left scale) and mean fluorescence intensity (black symbols, right scale). Oligonucleotide uptake into monocytes is inhibited by increasing concentrations of EDTA as well as heparin. Similar results were obtained for B-lymphocytes and granulocytes. Results are shown as means of experimental duplicates.

To reduce activation of coagulation, gain blood by intravenous puncture with a large volume butterfly-needle and fill directly into anticoagulant containing polypropylene tubes without drawing into a syringe. Anticoagulate with 0.8 mM EDTA or 1 IU/ml heparin. These are the minimal concentrations that guarantee complete anticoagulation. Proceed quickly with the remaining incubation steps as described below.

Oligonucleotides and cationic lipid reagents

Oligonucleotides can be purchased fluorescence-labeled at the 5´end. Fluoresceinisothiocyanat (FITC) or rhodamine can be used as fluorescent label. Rhodamine is suitable for conventional and confocal fluorescence microscopy, since it does not bleach as easily as FITC. Rhodamine however can not be detected by flow cytometry solely equipped with an argon ion laser. In our experiments on cellular oligonucleotide uptake, we used completely phosphorothioate-modified 18-mer oligonucleotides (5´ CAT GCT TTC AGT GCT CAT 3´ or 5´ CTA GGT TTG TCA CCT CTA 3´) at concentrations ranging from 0.125 to 1 µM.

For delivery of oligonucleotides into cells, the cationic lipid lipofectin (Gibco BRL, Eggenstein, Germany) is widely used. Lipofectin consists of equal parts of DOTMA (N-[1-(2,3-dioleyloxy)propyl]-N,N,N-trimethylammonium chloride; monovalent cationic lipid) and DOPE (dioleoyl phosphotidylethanolamine; neutral charge), and forms complexes with the polyanionic oligonucleotides. The ratio of positive and negative molar charge equivalents within these complexes is crucial for effective delivery of oligonucleotides to their target site (Felgner et al., 1993; Gershon et al., 1993). As stated by Felgner et al. (Felgner et al., 1993), the optimal charge ratio needs to be determined for each cell type and incubation condition used. The ratio of positive to negative molar charge equivalents can be calculated after Felgner et al (1993): The stock solution of 1 mg/ml lipofectin solution contains 0.5 mg/ml DOTMA (0.75 mM). Each DOTMA molecule contributes 1 positive charge equivalent. For example, 25 µg/ml lipofectin contain 18.8 µM positive charges. A concentration of 1 µM 18-mer oligonucleotide contains 17 µM negatively charged subunits (17 charges per oligonucleotide molecule). In this case, the ratio of positive to negative charges is 18.8 / 17 = 1.1. We found that a +/- ratio of 1.1 in the complexes is optimal for oligonucleotide uptake in human PBMC (see fig. 2) (Hartmann et al., 1998b). Antisense mediated suppression of tumor-necrosis-factor-α-synthesis using the same oligonucleotide-sequence was maximal at a similar +/- charge ratio of 0.9 (Hartmann et al., 1996).

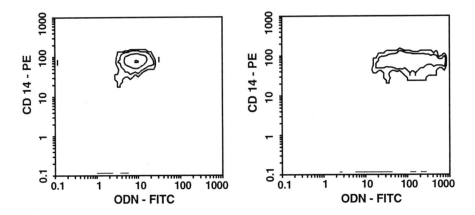

Figure 2: Enhancement of oligonucleotide uptake in monocytes by lipofectin: Mononuclear cells were incubated with 1 μM FITC labeled oligonucleotides alone (left) or together with 25 μg/ml lipofectin (+/- charge ratio 1.1) for 2 hours. Monocytes were identified by staining with PE-conjugated anti-CD14 monoclonal antibody. The mean fluorescence intensity (MFI) for oligonucleotide-FITC shifts from 9 (without lipofectin) to 133 (with lipofectin).

Complex formation of oligonucleotides and lipofectin is achieved as follows:

Incubate a 20-fold concentrated solution of lipofectin in sterile distilled H_2O in a polystyrol tube for about 30 minutes at room temperature. Add the same volume of a 20-fold-concentrated solution of oligonucleotide in sterile distilled H_2O. Mix gently by pipetting up and down, and allow complex formation for 30 min. at room temperature protected from light.

Incubation of cells with oligonucleotides

For flow cytometric analysis single cell suspensions are required. Monocytes are known to adhere to plastic surfaces (especially polystyrole). Removing adherent cells by scraping or trypsination may cause viability changes or denaturation of cell surface antigens. To avoid adherence of monocytes, we chose to incubate freshly prepared PBMC or whole blood samples in closed polypropylene tubes under rotation in a hybridization oven at 37°C. Culture medium was equilibrated with 5% CO_2 before use. Following the incubation of cells, the media is removed by centrifugation, and 200 μl cold PBS/BSA (1 %) containing 0.1 μM SdC28 (28mer poly cytidin phosphorothioate oligonucleotide) is added to each tube. SdC28 competes for polyanion binding sites on the cell surface and effectively removes all surface bound fluorescein-labeled oligonucleotide (Benimetskaya et al., 1997; Khaled et al., 1996). The use of fluorescein-labeled oligonucleoitdes for uptake studies may lead to artifactual results, since the extinction coefficient of fluorescein varies with the pH. This problem can be avoided by adding monensin to the cell preparations at a final concentration of 20 μM prior to flow cytometry or microscopy. Note, that the presence of monensin is not required during incubation of the cells with oligonucleotides.

Incubation of PBMC

Fill 200 µl aliquots of the prepared PBMC-suspension (cell density 10 x 10^6/ml) into 6 ml polypropylene tubes (Becton Dickinson). Dilute oligonucleotides or the preincubated lipofectin-oligonucleotide solution to twice the final concentration and add 200 µl per sample (final cell density 5 x 10^6/ml) As a control sample, add unconjugated fluorescein or fluorescein preincubated with lipofectin to cells. Close tubes and incubate for two hours at 37°C in a light-protected rotating hybridization oven. Keep cells in pellets on ice until analysis or further staining with antibodies against surface molecules.

Incubation of whole blood

For examination of cellular oligonucleotide uptake in whole blood, aliquot the prepared 10-fold concentrated lipofectin-oligonucleotide-solution or fluorescein preincubated with lipofectin (as a negative control) into polyproyplene tubes (10 µl each), add 90 µl of anticoagulated whole blood per sample and incubate the samples as described above for PBMC. After the last washing step, add 2ml of lysing-reagent (e.g. Ortho-mune™, Ortho Diagnostic Systems) and incubate at room temperature for 5 - 10 min. to lyse erythrocytes. Spin down cells, aspirate supernatant and wash once in PBS to remove lysed erythrocytes. Keep cells in pellets on ice until analysis or further staining with antibodies against surface markers.

Staining of non-viable cells and leukocyte subsets

Detection of non-viable cells by trypan blue exclusion technique using light microscopy is useful to quantify the percentage of dead cells. After the staining procedure, less than 5% of the cells should be found non-viable. The most commonly used dye to exclude non-viable cells in flow cytometry is propidium iodide (PI). It stains dead cells by intercalating in the DNA helix, whereas it does not enter viable cells with an intact membrane. PI fluoresces strongly orange-red and is detected in the fluorescence 2 or fluorescence 3 channel.

For differentiation of leukocyte subsets, antibodies against the CD markers on the cell surface are used. These antibodies have to be labeled with phycoerythrin (PE) or a tandem conjugate for the third color, e.g. phycoerythrin-Cy5 (PE Cy5), when FITC labeled oligonucleotides are used. The staining procedure is performed according to the manufacturer's instructions after the last washing step of the incubation with FITC labeled oligonucleotides.

Fluorescence detection

As described above, the use of fluorescence-labeled oligonucleotides allows to analyze two different parameters of antisense oligonucleotide uptake into blood cells: quantification of the fluorescence signal by flow cytometry gives us insights into the amount of oligonucleotides incorporated on a single cell basis, whereas fluorescence microscopy elucidates the distribution of the fluorescence signal in the cell. Therefore, both methods are complementary, and it is highly recommended to use them both for a better understanding of oligonucleotide uptake. For example, only flow cytometry allows a precise quantification of the improvement of antisense oligonucleotide uptake by cationic lipids, whereas it gives no insights into the improved translocation of oligonucleotides into the nucleus. The following chapters describing flow cytometry and fluorescence microscopy do not cover the whole field of fluorescence technology - many comprehensive reviews and textbooks are available for readers interested in more detailed information. Our aim is to describe protocols used in our lab for characterization of antisense uptake and to focus on problems you have to be aware when using fluorescence technology in antisense research.

Flow cytometry

We performed our experiments on both Becton Dickenson and Coulter flow cytometers, and other groups prefer cytometers from other companies. Based on our experience, there is no antisense oligonucleotide research-specific reason to choose one or the other. Today, the standard flow cytometer is equipped with one Argon laser, emitting light at 488 nm, and is able to record at least 5 parameters: forwardscatter (which represents cell size), sidescatter (which represents cell granularity) and three different fluorescence signals.

Instrument settings

Requirements for instrument settings are depending on the cell type you use and on the question you want to answer: obviously, forward- and sidescatter have to be adjusted for different cell lines, but in our experiments, PBMCs, whole blood and some lymphoblastic cell lines showed very similar morphological characteristics.

We are using FITC labeled antisense oligonucleotides and PE and/or PECy5 labeled monoclonal antibodies for surface staining of CD markers. The photomultiplier (PMT) settings for fluorescence measurement have to be optimized for every experiment. Enhancing the uptake of FITC labeled oligonucleotides by lipofectin, for example, leads to a very strong signal. Therefore, you should decrease the amplification of the FITC signal for this type of experiment. If not, you will run into problems since most of the events being measured are outside the linear working range of your cytometer on the upper end of the scale. This is extremely important if you are interested in the shift in mean fluorescence intensity (MFI) caused by cationic lipids or other substances: usually, the linear working range covers only the second and third decade of the logarithmic fluorescence intensity scale. Thus, only in these decades can a shift in MFI be correctly quantified, whereas the first and the fourth decade should be used for counting of „percent positive cells“. To guarantee a constant measurement of fluorescence intensity throughout a series of experiments, calibration beads have to be used daily (e.g., ImmunoCheck, Coulter). In addition, beads for controlling the linearity of fluorescence measurement are provided by different companies.

For multicolor analysis, compensation settings must be undertaken. This means an electronic correction of the spectral overlap from one fluorescence channel to the other. Compensation is a very controversially discussed topic in flow cytometry, and it is worth spending some time learning correct compensation before starting flow cytometric experiments. In general, correct compensation is more important for MFI analyses than for cell enumeration. Using the appropriate negative controls, discrimination between positive and negative cells is easy in most cases. However, when investigating antisense oligonucleotide uptake, we are more interested in quantification of fluorescence signals or analysis of changes in the MFI. Therefore, we have to ascertain that the MFI of the FITC signal is not disturbed by overlapping fluorescence of the CD marker used. On the other hand, when using cationic lipids, which generate a very strong FITC signal, it may be difficult to measure a CD marker with a weak expression in the PE channel. In this case, the use of PECy5-labeled CD antibodies is recommended. For appropriate compensation settings and plausibility control, it is necessary to prepare the cells with single color staining (only FITC-oligonucleotides on the one hand, only PE / PECy5 labeled CD marker on the other) in addition to the two or three color samples.

Data analysis

As mentioned above, there are two types of parameters you can analyze: First, you can enumerate the events stained positive by a certain marker. For example you can analyze the percentage of cells belonging to a distinct leukocyte subpopulation stained positive for FITC-labeled antisense oligonucleotides. From a theoretical point of view, and, in analogy to the nonspecific isotype controls used for antibodies against surface molecules, the negative control for a specific antisense oligonucleotide should be an a nonsense oligonucleotide labeled with the same fluorescence dye and inable to bind to the cells. However, such a negative control does not exist because the uptake of oligonucleotides is independent of the sequence. As confirmed by fluorescence microscopy, incorporation of oligonucleotides differs extremely between leukocyte subpopulations (Hartmann et al., 1998b). In our opinion, quantification of these differences and analysis of uptake enhancement is more important than the definition of a „really negative cell". Therefore, we chose the nonspecific background fluorescence of cells incubated with FITC alone or FITC and lipofectin as a negative control. Cells with a brighter fluorescence than these negative control samples are „positive" for oligonucleotide incorporation. Improving the uptake of oligonucleotides by cationic lipids may enhance the percentage of positive cells, which is correct for T cells and natural killer (NK) cells. However, nearly all B cells and monocytes are stained positive by FITC labeled oligonucleotides, even in the absence of cationic lipids. In these cells, further enhancement of uptake is only reflected by the amount of oligonucleotides per cell, which is represented by an increase in the mean fluorescence intensity (MFI) of the cell population (see fig. 2).

Fluorescence microscopy

Conventional fluorescence microscopy

Fluorescence microscopy is a necessary supplement to flow cytometric analysis, as it reveals the localization of detected fluorescence signals. Cell surface bound fluorescent oligonucleotides or artifacts caused by aggregates of fluorescent particles can thus be identified. Conventional fluorescence microscopy can be performed on vital cells in suspension after preparing them for flow cytometry.

After the last washing step, aspirate the supernatant except for 100 μl. Vortex, place a drop of this concentrated cell suspension on a chambered coverslide (Nunc Inc., Naperville, USA) and add 400 μl of PBS. Examine cells with an inverted fluorescence-microscope equipped with a high pressure mercury lamp and filters for FITC and rhodamine. Also view cells with phase contrast microscopy to assign fluorescence signals to cells and ensure their morphologic integrity. Phase contrast microscopy also allows approximate differentiation of monocytes and lymphocytes by size and appearance. Use rhodamine-conjugated oligonucleotides for photography because of FITC fades more quickly.

Confocal fluorescence microscopy

The confocal laser scanning fluorescence microscope is equipped with several lasers for excitation of fluorescent dyes. The argon ion laser (488 nm) is used to excite FITC, whereas the helium-neon-laser (543 nm or 633 nm) is needed for excitation of rhodamine or phycoerythrin. The laser beam is focused on a defined level of the microscopic sample. This level of the sample is scanned by the laser which excites the fluorescent dye. Fluorescence signals are distinguished by band pass filters, detected and processed to obtain two-dimensional digital images. Successive optical sections from the top to the bottom of cells (vertical distance 500 nm) can be obtained using this technique. This enables differentiation between intracellular and extracellularly bound fluorescence, and localization of fluorescence signals to intracellular structures. Specific cell types can be identified by staining with specific antibodies conjugated to a second fluorescent dye. In the same cell, the fluorescence pattern of both fluorescent dyes is detected and processed to obtain two color images.

Movement of viable cells during the scanning process may complicate the acquisition of confocal images. However, fixation of the cells changes the intracellular localization of oligonucleotides and should be avoided.

3.3 CONCLUSIONS

The variance in antisense oligonucleotide uptake in different cell types influences the efficacy of these therapeutic agents. Flow cytometry enables simultaneous analysis of both oligonucleotide uptake and surface membrane markers on a single cell basis. Therefore, it is a suitable method for investigating the possible usefulness of antisense strategies in different specific cell types. In addition, the effect of different agents tested for enhancement of olgonucleotide uptake, like cationic lipids, can be studied on a semiquantitative basis. However, the efficacy of antisense oligonucleotides depends not only on cellular uptake, but also on the distribution within the cell. As described in this chapter, fluorescence microscopy is necessary to complement flow cytometry, because of the additional information on subcellular distribution of fluorescence signals.

Despite of the above-mentioned influence of activation by adhesion, endotoxin contamination, or by the uptake inhibiting effect of anticoagulation, we want to point to the competition for cellular binding sites between oligonucleotides and other substances, such as fibrinogen (Benimetskaya et al., 1997; Khaled et al., 1996). We have to assume a significant impact of these factors present in the *in vivo*, but not automatically in the *in vitro* situation, not only on spontaneous, but also on cationic lipid-mediated uptake of oligonucleotides. Therefore, studies on antisense oligonucleotide uptake *in vitro* should be interpreted carefully regarding their relevance for the *in vivo* situation (Hartmann et al., 1997).

3.4 REFERENCES

Albrecht T, Schwab R, Peschel C, Engels HJ, Fischer T, Huber C, Aulitzky WE. Cationic lipid mediated transfer of c-abl and bcr antisense oligonucleotides to immature normal myeloid cells: uptake, biological effects and modulation of gene expression. *Ann Hematol* 1996;72:73-79

Beltinger C, Saragovi HU, Smith RM, Lesauteur L, Shah N, Dedionisio L, Christensen L, Raible A, Jarett L, Gewirtz AM. Binding, uptake, and intracellular trafficking of phosphorothioate-modified oligodeoxynucleotides. *J Clin Invest* 1995;95:1814-23

Benimetskaya L, Loike JD, Khaled Z, Loike G, Silverstein SC, Cao L, el-Khoury J, Cai TQ, Stein CA. Mac-1 (CD11b/CD18) is an oligodeoxynucleotide-binding protein. *Nat Med* 1997;3:414-20

Bennett CF, Chiang MY, Chan H, Shoemaker JE, Mirabelli CK. Cationic lipids enhance cellular uptake and activity of phosphorothioate antisense oligonucleotides. *Mol Pharmacol* 1992;41:1023-33.

Bennett CF, Kornbrust D, Henry S, Stecker K, Howard R, Cooper S, Dutson S, Hall W, Jacoby HI. An ICAM-1 antisense oligonucleotide prevents and reverses dextran sulfate sodium-induced colitis in mice. *J Pharmacol Exp Ther* 1997;280:988-1000

Bradbury J. Antisense drugs move towards the clinic. *Lancet* 1997;349:259

Chavany C, Connell Y, Neckers L. Contribution of sequence and phosphorothioate content to inhibition of cell growth and adhesion caused by c-myc antisense oligomers. *Mol Pharmacol* 1995;48:738-46

Felgner J, Bennet F, Felgner PF. Cationic lipid-mediated delivery of polynucleotides. *Meth Compan Meth Enzymol* 1993;5:67-75

Fisher TL, Terhorst T, Cao X, Wagner RW. Intracellular disposition and metabolism of fluorescently-labeled unmodified and modified oligonucleotides microinjected into mammalian cells. *Nucleic Acids Res* 1993;21:3857-65

Gershon H, Ghirlando R, Guttman SB, Minsky A. Mode of formation and structural features of DNA-cationic liposome complexes used for transfection. *Biochemistry* 1993;32:7143-51.

Hartmann G, Bidlingmaier M, Jahrsdörfer B, Endres S. Oligonucleotides: Extrapolating from in vitro to in vivo. *Nat Med* 1997;3:702

Hartmann G, Bidlingmaier M, Jahrsdörfer B, Krug A, Hacker U, Eigler A, Endres S. Oligonucleotide uptake in leucocytes is dependent on extracellular calcium: a hint for the involvement of adhesion molecules? *Nucleos Nucleotid* 1998a;in press

Hartmann G, Krug A, Bidlingmaier M, Hacker U, Eigler A, Albrecht R, Strasburger C, Endres S. Spontaneous and cationic lipid-mediated uptake of antisense oligonucleotides in human monocytes and lymphocytes. *J Pharmacol Exp Ther* 1998b ;in press

Hartmann G, Krug A, Eigler A, Moeller J, Murphy J, Albrecht R, Endres S. Specific suppression of human tumor necrosis factor-α synthesis by antisense oligodeoxynucleotides. *Antisense Nucl Acid Drug Dev* 1996a;6:291-99

Hartmann G, Krug A, Waller-Fontaine K, Endres S. Oligodeoxynucleotides enhance LPS-stimulated synthesis of tumor necrosis factor: dependence on phosphorothioate modification and reversal by heparin. *Mol Med* 1996b;2:429-38

Iversen PL, Zhu S, Meyer A, Zon G. Cellular uptake and subcellular distribution of phosphorothioate oligonucleotides into cultured cells. *Antisense Res Dev* 1992;2:211-22

Keller ET, Ershler WB. Effect of IL-6 receptor antisense oligodeoxynucleotide on in vitro proliferation of myeloma cells. *J Immunol* 1995;154:4091-98

Khaled Z, Benimetskaya L, Zeltser R, Khan T, Sharma HW, Narayanan R, Stein CA (1996) Multiple mechanisms may contribute to the cellular anti-adhesive effects of phosphorothioate oligodeoxynucleotides. Nucleic Acids Res 24:737-745.

Krieg AM, Gmelig MF, Gourley MF, Kisch WJ, Chrisey LA, Steinberg AD. Uptake of oligodeoxyribonucleotides by lymphoid cells is heterogeneous and inducible. *Antisense Res Dev* 1991;1:161-71

Loke SL, Stein CA, Zhang XH, Mori K, Nakanishi M, Subasinghe C, Cohen JS, Neckers LM. Characterization of oligonucleotide transport into living cells. *Proc Natl Acad Sci USA* 1989;86:3474-78

Lund OS, Nielsen JO, Hansen JES. Inhibition of HIV-1 in vitro by c-5 propyne phosphorothioate antisense to rev. *Antivir Res* 1995;28:81-91

Marti G, Egan W, Noguchi P, Zon G, Matsukura M, Broder S. Oligodeoxyribonucleotide phosphorothioate fluxes and localization in hematopoietic cells. *Antisense Res Dev* 1992;2:27-39

Matteucci MD, Wagner RW. In pursuit of antisense. *Nature* 1996;384:20-22

Peng BH, Mehta NH, Fernandes K, Chou CC, Raveche E. Growth inhibition of malignant CD5+B (B-1) cells by antisense IL-10 oligonucleotide. *Leukemia Res* 1995;19:159-67

Pirruccello SJ, Perry GA, Bock PJ, Lang MS, Noel SM, Zon G, Iversen PL. HIV-1 rev antisense phosphorothioate oligonucleotide binding to human mononuclear cells is cell type specific and inducible. *Antisense Res Dev* 1994;4:285-89

Reed JC, Stein C, Subasinghe C, Haldar S, Croce CM, Yum S, Cohen J. Antisense mediated inhibition of BCL2 protooncogene expression and leukemic cell growth and survival: comparisons of phosphodiester and phosphorothioate oligodeoxynucleotides. *Cancer Res* 1990;50:6565-70

Srinivasan SK, Tewary HK, Iversen PL. Characterization of binding sites, extent of binding, and drug interactions of oligonucleotides with albumin. *Antisense Res Dev* 1995;5:131-39

Temsamani J, Kubert M, Agrawal S. A rapid method for quantitation of oligodeoxynucleotide phosphorothioates in biological fluids and tissues. *Anal Biochem* 1993;215:54-58

Temsamani J, Kubert M, Tang J, Padmapriya A, Agrawal S. Cellular uptake of oligodeoxynucleotide phosphorothioates and their analogs. *Antisense Res Dev* 1994;4:35-42

Vaughn JP, Iglehart JD, Demirdji S, Davis P, Babiss LE, Caruthers MH, Marks JR. Antisense DNA downregulation of the ERBB2 oncogene measured by a flow cytometric assay. *Proc Natl Acad Sci USA* 1995;92:8338-42

Wagner RW, Matteucci MD, Lewis JG, Gutierrez AJ, Moulds C, Froehler BC. Antisense gene inhibition by oligonucleotides containing C-5 propyne pyrimidines. *Science* 1993;260:1510-11

Wang S, Lee RJ, Cauchon G, Gorenstein DG, Low PS. Delivery of antisense oligodeoxyribonucleotides against the human epidermal growth factor receptor into cultured KB cells with liposomes conjugated to folate via polyethylene glycol. *Proc Natl Acad Sci USA* 1995;2:318-22

Webb A, Cunningham D, Cotter F, di Clarke PA, SF, Ross P, Corbo M, Dziewanowska Z. BCL-2 antisense therapy in patients with non-Hodgkin lymphoma. *Lancet* 1997;349:1137-41

Yahata N, Kawai S, Higaki M, Mizushima Y. Antisense phosphorothioate oligonucleotide inhibits interleukin 1 beta production in the human macrophage-like cell line, U937. *Antisense Nucleic Acid Drug Dev* 1996;6:55-61

Zhao Q, Matson S, Herrera CJ, Fisher E, Yu H, Krieg AM. Comparison of cellular binding and uptake of antisense phosphodiester, phosphorothioate, and mixed phosphorothioate and methylphosphonate oligonucleotides. *Antisense Res Dev* 1993;3:53-66

Zhao Q, Song X, Waldschmidt T, Fisher E, Krieg AM. Oligonucleotide uptake in human hematopoietic cells is increased in leukemia and is related to cellular activation. *Blood* 1996;88:1788-95

Zhao Q, Waldschmidt T, Fisher E, Herrera CJ, Krieg AM. Stage-specific oligonucleotide uptake in murine bone marrow B-cell precursors. *Blood* 1994;84:3660-66

4 Strategies for Targeted Uptake of Antisense Oligonucleotides in Hepatocytes

Wolf-Bernhard Offensperger, Jerzy Madon*, Christian Thoma, Darius Moradpour, Fritz von Weizsäcker, Hubert E. Blum

Department of Medicine II, University of Freiburg, Germany
*Department of Medicine, University Hospital Zuerich, Switzerland

4.1 SUMMARY

Antisense oligonucleotides have shown great efficacy in the selective inhibition of gene expression. In hepatology, the interest in antisense oligonucleotides concentrates on their use in infections caused by the hepatitis B virus (HBV) or the hepatitis C virus (HCV). Significant inhibition of HBV replication could be demonstrated *in vitro* in hepatoma cells and in primary hepatocytes as well as *in vivo* in animal models. To deliver oligonucleotides to hepatocytes several strategies were explored including receptor-mediated endocytosis *via* the asialoglycoprotein-receptor. In this chapter the construction of conjugates consisting of an asialoglycoprotein carrier molecule, poly-L-lysine, DNA and replication-defective adenovirus dl312 as endosomolytic agent is described. Using this delivery system the effective uptake of oligonucleotides into hepatoma cells and into duck liver *in vivo* could be achieved.

4.2 INTRODUCTION

Hepatic gene therapy

Genetically, human diseases can be classified into the following three major categories:
1. Monogenetic diseases that are caused by a single gene defect.
2. Complex genetic diseases that are associated with mutations of several genes, only some of which have been identified to date. Several common human diseases belong to this category, such as most malignancies, diseases of the cardiovascular system, hypertension, arthritis, diabetes, and others.
3. Acquired genetic diseases that include all infections as well as some malignancies that are associated with genetic alterations that are frequently accumulated during a lifetime.

Examples of monogenetic diseases are familial hypercholesterolemia, caused by a defect in the low density lipoprotein receptor, phenylketonuria, which results from mutations in the phenylalanine hydroxylase gene, hyperammonemia, caused by inherited defects in the urea cycle, hemophilia B, resulting from factor IX deficiency, lysosomal storage diseases and Wilson's disease (Ledley, 1993; Chang et al., 1994; Alt et al., 1995). For some of these diseases, a possible treatment is the orthotopic liver transplantation, which, however, is associated with considerable morbidity and mortality. An alternative potential therapeutic approach is the transfer of therapeutic nucleic acids into the affected organs. Because there is no reliable and safe technique allowing the site-specific integration of DNA into the human genome available at the moment, the replacement of the defective gene by a copy of the functional gene is not possible so far. This is the reason why almost all methods of gene therapy focus on transfering the therapeutic gene into somatic cells without replacing the abnormal gene. Of the available methods of gene delivery, viruses have proved the most efficient so far. The objective of virus-mediated gene transfer is to take advantage of the highly evolved processes by which viruses normally enter cells and alter the characteristics of the infected cell by their own genetic material. There is now extensive experience with retroviruses whose main advantages include their small size and their ease of manipulation with stable colinear integration into the host genome. Currently, alternative viral vectors with potential advantages over retroviruses in specific applications are being studied. Adenoviruses can infect non-dividing cells, can be concentrated to high titres, and are comparatively high efficiency vectors. Adeno-associated viruses are ubiquitous and non-pathogenic in humans and can infect also non-replicating cells, but, like retroviruses and adenoviruses, are limited in the size of the foreign gene that can be inserted.

For diseases caused by the expression of acquired genes, such as viral genes, blocking of gene expression can be an effective therapeutic approach (Von Weizsäcker et al., 1997). Several strategies can be employed: interfering with the transcription of genes by binding of single-stranded nucleic acids to double-stranded DNA, forming a triple helix structure, by hybridization of RNA molecules possessing endoribonuclease activity (ribozymes) to target RNA molecules, resulting in sequence-specific RNA cleavage, by blocking translation through binding of antisense oligodeoxynucleotides (oligonucleotide) to RNA, and by intracellular synthesis of peptides or proteins that interfere with their normal counterpart (dominant negative mutant strategy).

Mainly two viruses are involved in chronic liver disease, namely HBV and HCV. Infection with HBV is endemic throughout much of the world with an estimated 400 million persistently infected people (Lee, 1997). HBV infection is associated with a wide spectrum of clinical presentations, ranging from the healthy carrier state to acute/fulminant or chronic hepatitis and liver cirrhosis. Furthermore, chronic HBV infection clearly contributes to the development of hepatocellular carcinoma, worldwide a leading cause of death from cancer. While interferon-alpha has been used to treat chronic hepatitis, it has been only partially successful in elimination of chronic infection (Hoofnagle et al., 1997). While other potential therapeutics such as lamivudine are in clinical evaluation, the need for alternative therapeutic strategies has provided impetus to develop novel concepts. Several groups have reported inhibition of HBV gene expression and replication with antisense oligonucleotides (Goodarzi et al., 1990; Blum et al., 1991). A series of oligonucleotides against HBV were evaluated using a hepatocellular carcinoma-derived cell line stably transfected with HBV DNA (Hep 2.2.15) (Korba et al., 1995). The first *in vivo* studies using oligodeoxynucleotides in hepadnaviral infection were done in the Pekin duck model (Offensperger et al., 1993). Duck hepatitis B virus (DHBV) is a member of the hepadnavirus family. This family includes several viruses which have been studied in detail at the molecular level. The prototype of this family is HBV. Closely related viruses have been identified in woodchucks, ground squirrels and in the Pekin duck. Initially, 9 different phosphorothioate-modified antisense oligonucleotides (16-18 nucleotides in length) directed against different regions of the DHBV genome were tested *in vitro* in primary duck hepatocytes infected by DHBV. The most effective antisense oligonucleotide was directed against the 5'-region of the preS gene and resulted in greater than 90% inhibition of viral replication and gene expression *in vitro*. A lack of this antiviral activity by 'sense oligonucleotides' indicated that the oligonucleotides were acting in a polarity-specific manner. These *in vitro* analyses were followed by *in vivo* treatment of DHBV-infected Pekin ducks with this antisense oligonucleotide. Analysis of viral DNA in the liver revealed a dose-dependent inhibition of viral replication with nearly complete elimination of replicating viral DNA at a daily dose of 20 µg/g body weight. In addition, effective inhibition of viral gene expression could be demonstrated. Clinical side effects of this *in vivo* therapy were excluded by the determination of several clinico-chemical parameters, including alanine aminotransferase, aspartate aminotransferase, cholinesterase, total protein and albumin concentration. More recently, studies in nude mice (Yao et al., 1996) and the woodchuck hepatitis virus model of HBV infection (Bertholomew et al., 1995) showed the *in vivo* applicability of this approach.

4.3 OLIGONUCLEOTIDES: UPTAKE, TRAFFICKING AND TARGETING

Antisense oligonucleotides have shown great efficacy in the selective inhibition of gene expression (Wagner, 1994). However, the therapeutic applications of such antisense oligonucleotides are currently limited by three problems: 1) low stability under physiological conditions, 2) poor cellular uptake and trafficking and 3) lack of tissue specificity.

1. The instability problems have been largely overcome by using backbone-modified oligonucleotides that are more resistant to nucleases. Phosphorothioate-modified or methylphosphonate-modified oligonucleotides are more resistant to enzymatic digestion than the corresponding unmodified oligonucleotides (Szymkowski, 1996).

2. Problems with the cellular uptake of antisense oligonucleotides have been more difficult to solve. Phosphodiester oligonucleotides and the widely used phosphorothioate-modified oligonucleotides, which contain a single sulfur substituting for oxygen at a nonbridging position at each phosphorus atom, are polyanions. Accordingly, they possess little or no ability to diffuse across cell membranes and are taken up by cells only through energy-dependent mechanisms. This appears to be accomplished primarily through a combination of adsorptive endocytosis and fluid-phase endocytosis, which may be triggered in part by the binding of the oligonucleotide to receptor-like proteins present on the surface of a wide variety of cells (Gewirtz et al., 1996). After internalization, confocal and electron microscopy studies have indicated that most of the oligonucleotides enter the endosome/lysosome compartment. These vesicular structures become acidified and acquire enzymes that degrade the oligonucleotides. Some of the oligonucleotides escape from the vesicles intact, enter the cytoplasm, and then diffuse into the nucleus where they presumably bind to their mRNA or gene target. The processes that control release from the vesicles and regulate trafficking between the cytoplasm and the nucleus are not well understood. To resolve the problems of uptake and trafficking of oligonucleotides, a large number of strategies have been tested in order to augment the rate of cellular internalization and to increase the rate at which they pass through the endosomal membrane. Macromolecular materials in which the oligonucleotide is dissolved, entrapped or encapsulated or to which the oligonucleotide is adsorbed or attached are currently under investigation. Delivery vehicles which were tested include cationic lipids, liposomes, lipopolyamines, conjugation to fusogenic peptides or cholesterol.

   Cationic lipids form complexes with DNA through charge interactions (Gao et al., 1995). DNA in the form of a complex is protected from degradation. DNA-lipid complexes bind to the negatively charged cell surfaces due to the presence of excess positive charges in the complex. The nature of nonspecific interaction results in efficient transfection of many cell types (Bennett et al., 1992). Relatively high transfection efficiency comes from the intrinsic membrane rupturing capability of cationic liposomes as a result of destabilizing the endosome and/or plasma membrane. However, cationic lipids such as lipofectin cause non-specific cytotoxic effects if applied alone or together with antisense oligonucleotides (Ledley, 1995). Only recently new generation compounds with significantly less cytotoxic effects have been developed. These new compounds may become interesting tools for antisense delivery. Nevertheless, the benefit of cationic lipids for *in vivo* applications remains unclear since improved uptake and antisense inhibition in cell culture are not necessarily observed in laboratory animals. The cationic polymer polyethylenimine could be shown to be a highly efficient vector for delivering oligonucleotides both *in vitro* and *in vivo* (Boussif et al., 1995; Chemin et al., Submitted). For conventional liposomes an enhanced cellular uptake has been described in cell culture (Wang et al., 1987). Significant improvements of oligonucleotide transfection were reported using a fusogenic peptide, derived from the influenza hemagglutinin envelope protein (Bongartz et al., 1994). This peptide changes conformation at low pH and destabilizes the endosomal membranes thus resulting in increased cytoplasmic gene delivery. In a further study oligonucleotides were synthesized with a cholesteryl group tethered at the 3'-terminal internucleoside link (Letsinger et al., 1989). This modification, introduced to enhance interaction of the polyanions with cell membranes, significantly increased the antiviral activity of the oligomer.

3.   Oligonucleotide targeting implies the manipulation of distribution of oligonucleotides in the whole body: more tissue-selective localization is obtained by coupling of the oligonucleotide to a carrier specifically recognized by a receptor or carrier protein on the surface of the cell type targeted. Ideally, through association of the targeted oligonucleotide to such a 'homing device', distribution in the organism is dictated by the chosen carrier, no longer by the physicochemical properties of the oligonucleotide itself. The obvious aim is to increase therapeutic concentrations in particular cells (active targeting), to prevent distribution to tissues where toxicity of the particular drug is produced (passive targeting) or both. Targeting vehicles include antibodies and ligands. Liposomes may be linked to specific antibodies to achieve specific tissue targeting (immunoliposomes). Recently, a monoclonal antibody (AF-20) against the human hepatocellular carcinoma cell line FOCUS was produced followed by the production of immunoliposomes *via* covalently coupling AF-20 to liposomes, resulting in the interaction of these immunoliposomes with and enhanced gene delivery into hepatocellular carcinoma cell lines (Moradpour et al., 1995; Compagnon et al., 1997). Further work demonstrated that antibody-targeted liposomes containing oligonucleotides selectively inhibited the replication of vesicular stomatitis virus in mouse L929 cells (Leonetti et al., 1990).

The use of cell-specific ligands opens the possibility to specifically transduce the cell type which possesses the respective receptor. The principle of all receptor-mediated transport systems for oligonucleotides is to subvert the efficient cellular mechanisms of receptor-mediated internalization of proteins in such a way that ligands bound to DNA are recognized by the receptors and are carried efficiently across the plasma membrane into the cell. Efficient receptor-mediated DNA delivery was reported *via* the transferrin receptor (Cotten et al., 1993a), the folate receptor (Wang et al., 1995), the polymeric immunoglobulin receptor of the airway epithelium (Ferkol et al., 1995), the macrophage mannose receptor (Ferkol et al., 1996) and the asialoglycoprotein receptor (ASGP-R). The ASGP-R (Steer, 1996) is an integral membrane glycoprotein and is found exclusively on hepatocytes. It is distributed randomly along the sinusoidal and lateral plasma membrane with little evidence of expression along the canalicular membrane. The receptor mediates the specific recognition and uptake of glycoproteins with terminal galactose or N-acetylgalactosamine. More receptor-mediated processes have been identified on hepatocytes including receptors for LDL-particles, HDL-particles, apolipoprotein E (Rensen et al., 1995), epidermal growth factor, hemopexin, chylomicron remnants, hemoglobin, insulin and polymeric IgA/IgM (called secretory component) (Meijer et al., 1995). On Kupffer cells, receptors for immune and complement complexes (Fc and C3 receptors) and for fibronectin were detected, that are also able to mediate endocytosis of opsonized particulate material (Meijer et al., 1995). In addition, a galactose-particle receptor is described, recognizing cells and other particles that expose galactose groups. In contrast to the ASGP-R on hepatocytes, this galactose-specific receptor is highly size-sensitive. As in other macrophages, a mannose/N-acetylglucosamine/fucose receptor has been identified that not only recognizes terminal mannose and N-acetylglucosamine groups on glycoproteins, but may also accommodate glycoproteins with terminal glucose and fructose groups. Kupffer cells, along with hepatocytes, can also endocytose lysosomal proteases by complex formation with circulating alpha-2-macroglobulin. Finally, Kupffer cells exhibit receptors for positively charged as well as negatively charged proteins (scavenger receptors). Although the above-mentioned receptors in Kupffer cells were once supposed to be specific for this cell type, it is now agreed that many of them are also present on endothelial cells. The endothelial cells also have receptors for LDL particles. Scavenger receptors may even be predominantly localized in the endothelial cells (Kamps et al., 1997).

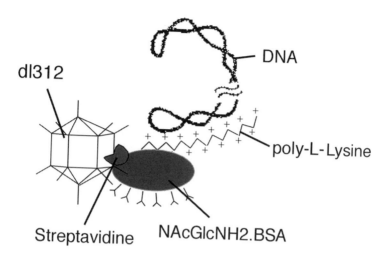

Figure 1: Delivery complex consisting of the adenovirus particle, the protein conjugate with the ligand and streptavidine, and the plasmid DNA or the oligonucleotides.

4.4 ADENOVIRUS ENHANCED, ASGP-R MEDIATED OLIGONUCLEOTIDE DELIVERY

A variety of agents can be delivered to hepatocytes using ASGPs as vehicles. Fiume and colleagues coupled the antiviral drug ara-AMP (9-ß-D-arabinofuranosyladenine 5'-monophosphate) to the carrier lactosaminated albumin (Fiume et al., 1981). These investigators made the interesting observation that after injection of the conjugate into mice and woodchucks, synthesis of viral DNA was inhibited at lower concentrations than it was with the uncoupled drug. Jansen and colleagues could demonstrate that human hepatocytes can carry out endocytosis of lactosaminated human serum albumin and the ara-AMP conjugate by means of the ASGP-R (Jansen et al., 1993). Previous data indicated that in patients with chronic hepatitis B ara-AMP coupled to lactosaminated human albumin could exert the antiviral activity of ara-AMP without producing the neurotoxic side effects that limits treatment with the free drug (Torrani et al., 1996).

In these studies, agents were covalently coupled to ASGPs. However, in the case of DNA or oligonucleotides, covalent coupling to the ASGP could result in damage to the bases and altered transcription. To overcome this problem, mainly two possibilities were investigated:

1. Incorporation of ASGPs into liposomes or lipopolyamines containing foreign DNA or oligonucleotides. For example, incorporation of lactosyl ceramide into liposome envelopes containing the preproinsulin I gene and intravenous injection of these modified liposomes resulted in a preferential delivery to the liver (Soriano et al., 1983). Furthermore, targeted gene transfer into hepatoma cells was achieved with lipopolyamine-condensed DNA particles presenting galactose ligands (Remy et al., 1995).

2. Construction of a targetable conjugate consisting of two components: an ASGP carrier molecule, covalently coupled to a polycation. The polycation, because of its positive charges, could bind the negative charges of DNA in a strong, but non-damaging electrostatic interaction yielding a soluble polycation-DNA complex. It could also serve to protect the DNA from nuclease attack. In their pioneering work Wu and coworkers coupled poly-L-lysine to an ASGP to form a targetable DNA carrier (Wu et al., 1987). Subsequent addition of DNA to form a soluble ASGP-poly-L-lysine-DNA complex could permit targeted delivery of DNA specifically to ASGP-R-bearing cells. The objective of further work of Wu and coworkers was to determine whether this targeted delivery system can be used to deliver single-stranded antisense oligonucleotides. A cell line, HepG2 2.2.15 that has the ASGP-R and is stably transfected with HBV, was exposed to complexed antisense oligonucleotides or control (Wu et al., 1992). The data indicated that antisense oligonucleotides complexed by a soluble DNA-carrier system can be targeted to cells via ASGP-R resulting in specific inhibition of HBV gene expression and replication. Further work studied the biodistribution of ASGP-poly-L-lysine-oligonucleotides after intravenous injection in rats and showed that the accumulation of the ASGP-poly-L-lysine-oligonucleotides complex greatly exceeded the accumulation of the antisense alone (Lu et al., 1994). *In vivo* gene delivery with ASGP-poly-L-lysine-DNA complexes has been described in several different experimental models. Initial studies demonstrated that the activity of chloramphenicol acetyltransferase could be constituted in the liver after gene transfer with such complexes (Wu et al., 1988). Subsequent studies have demonstrated expression of the LDL receptor in LDL-deficient rabbits (Wilson et al., 1989), of albumin in analbuminemic rats (Wu et al., 1991), and of methylmalonyl CoA mutase in mice (Stankovics et al., 1992). However, this delivery system turned out to have a low gene transfer efficiency with less than 0.1% of cells exposed to the complex expressing foreign DNA (Cristiano et al., 1993). Partial hepatectomy improved gene delivery by prolonged persistence of complexed DNA in cytoplasmic vesicles (Chowdhury et al., 1993). Condensing DNA with galactosylated poly-L-lysine by titration with NaCl resulted in complexes of defined size and shape and in specific targeting to the livers of the animals (Perales et al., 1994).

A significantly improved efficiency of gene delivery was obtained by including human adenoviral particles into the receptor-mediated gene delivery concept (Cristiano et al, 1993; Curiel et al., 1991; Wagner et al., 1992). The contribution of adenoviral particles to gene delivery efficiency is probably caused by the natural entry mechanism of adenoviruses that induce passage through or disruption of endosomal membranes. Initial experiments were performed with adenovirus dl312, a replication-defective Ela-deficient strain of the human adenovirus type 5 (Ad5). Although a DNA-transduction efficiency of more than 90% was achieved by adenovirus dl312, long-term cytopathic effects caused by replicational and/or transcriptional leakiness of the adenoviral mutant strain caused serious problems. To circumvent the toxicity of adenoviral particles, the use of adenoviruses from distant species seemed an attractive possibility. In this context, it has been shown that the avian adenovirus type 1 (chicken embryo lethal orphan virus) showed endosomolytic activity similar to that of human adenoviruses and was used successfully for receptor-mediated gene delivery into different mammalian cell types without long-term cytopathic effects (Cotten et al., 1993).

In the context of our interest in molecular therapeutic strategies against HBV infection in the animal model of DHBV infection the availability of efficient gene transfer methods into the duck liver was a precondition. Previous work of our group could establish a highly efficient receptor-mediated delivery system for DNA to avian liver cells (Madon et al., 1996). The delivery system consisted of adenovirus particles, protein conjugate and plasmid DNA. A schematic illustration of the delivery system is shown in Figure 1. The replication-defective mutant strain dl312 of the human serotype 5 was used as an endosome-disruption factor. The protein conjugate contained NAcGlcNH2.BSA (Lig) as the ligand for the avian liver ASGP (Mellow et al., 1988), poly-L-lysine, and streptavidin. For preparation of protein conjugates, a simple and reproducible coupling method was developed. After protein coupling, L-lysine was added in excess to block the remaining reactive groups. The conjugates were purified by cation exchange high-performance liquid chromatography. Fractions eluting between 1.45 mol/L and 2.1 mol/L of NaCl contained the correct conjugate. The plasmids pCMVluc or pCMVlacZ were used to study the efficiency. For *in vitro* studies the chicken hepatoma cell line LMH was used. This cell line has been used successfully for the study of hepadnaviral replication and gene expression (Condreay et al., 1990). Transfection of HBV or DHBV DNA into LMH cells resulted in high levels of viral gene expression and replication with formation and export of infectious virions. Reporter assays were performed 24 hours after the transfection of the LMH chicken hepatoma cells with the complexes. In each case, 1×10^7 pfu adenoviral particles and 200 ng pCMVluc DNA were used. The main characteristics of this system are summarized in Figure 2a. Incubation of the cells with pCMVluc alone, or with the combination of pCMVluc and the protein conjugate or the adenovirus alone did not result in detectable luciferase activity. Incubation of the cells with a complex which did not include the ligand did again not show measurable luciferase activity. Constructing a complex of adenovirus particles, poly-L-lysine, pCMVluc and BSA as unspecific ligand also did not lead to luciferase activity. However, when the delivery complex contained adenovirus particles, the protein conjugate with NAcGlcNH2.BSA and pCMVluc, very high levels of luciferase activity were observed. No cytopathic effects could be seen with any combination of the delivery complex components.

More than 95% of the LMH cells incubated with the complete delivery complex stained very strongly blue after treatment with X-gal. To determine whether the DNA delivery was mediated specifically through receptor-mediated endocytosis, the cells were incubated with the delivery complexes in the presence of increasing concentrations of free ligand NAcGlcNH2.BSA or the control protein BSA. As shown in Figure 2b, delivery of pCMVluc was strongly inhibited by NAcGlcNH2.BSA with no luciferase activity left at a 200-fold excess of the free ligand. No inhibition could be seen using BSA as competitor.

To determine the delivery efficiency of antisense oligonucleotides, a fluorescein-labeled oligonucleotide Enc-Fluo was used. This oligonucleotide was phosphorothioate modified and had the following sequence: 5' CAGTGGGACATGTACA 3'. LMH cells were incubated with delivery complexes containing this fluorescein-labeled oligonucleotide. While incubation with the noncomplexed fluorescein-labeled oligonucleotide added directly to the culture medium at a final concentration of 1.5 µmol/L resulted in only insignificant labeling of LMH cells (Figure 3), complexed fluorescein-labeled oligonucleotides could be detected in more than 95% of the LMH cells, indicating a high delivery efficiency. The highest intensity of fluorescence was observed at an oligonucleotide concentration of 3.0 µmol/L in the complex formation mixture. A further increase of the oligonucleotide concentration did not affect delivery efficiency. However, the amount of oligonucleotides delivered into the cells clearly depended on the time of incubation of the delivery complex with the LMH cells: longer incubation times resulted in greater oligonucleotide levels in the cells. The intracellular distribution of the delivered oligonucleotides was analyzed by confocal microscopy. The highest intensity of fluorescent oligonucleotides was found in the nuclei of the majority of cells.

In the next step fluorescein-labeled oligonucleotides (200 µg each) in complexed or non-complexed form were injected into the foot vein of 1 week-old Pekin ducklings (weight about 100 g). Two hours later, the ducklings were sacrificed. The livers were removed and digested by collagenase. The percentage of FITC-positive hepatocytes was determined by FACS-analysis. As is shown in Figure 4, the uptake of phosphorothioate-modified oligonucleotides in complexed or non-complexed form did not differ. However, unmodified oligonucleotides were not taken up into hepatocytes in non-complexed form, in complexed form their uptake was comparable to that of the phosphorothioate-modified oligonucleotides.

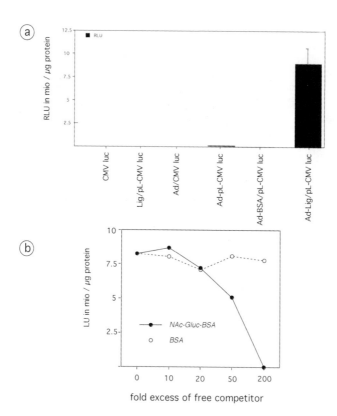

Figure 2: a) Characteristics of the delivery complex Ad-Lig/pL-CMVluc in duck hepatoma LMH cells. Determination of luciferase expression in RLU/ug protein.
b) Competition assays of Ad-Lig/pL-CMVluc delivery to LMH cells using the free ligand NacGlcNH2.BSA and the control protein BSA.

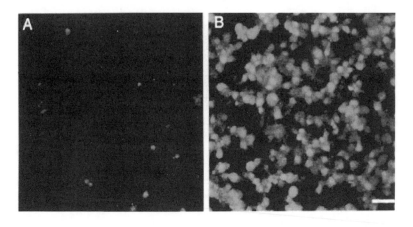

Figure 3: Delivery of antisense oligonucleotides to LMH cells (53). The cells were incubated for 8 hours with the oligonucleotides alone (A) or the delivery complex containing the oligonucleotide added directly to the culture medium (final concentration 1.5 µmol/L). After three washes with IMDM and with the addition of IMDM supplemented with 10% FCS, the cells were analyzed directly for fluorescence using a Leica confocal laser scanning microscope with long-distance objective of 20x/0.4 (Leica, Wetzlar, Germany).

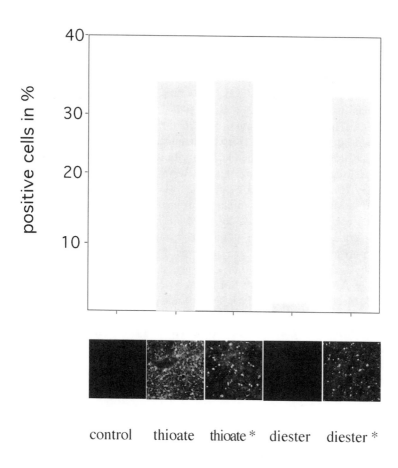

Figure 4: Uptake of fluorescein-labeled, phosphorothioate-modified (thioate) and unmodified (diester) oligonucleotides in complexed form (*) or in non-complexed form into the liver of Pekin ducklings. Determination of the percentage of FITC-positive hepatocytes by FACS-analysis.

4.5 PROTOCOLS

Preparation and purification of protein conjugates

The standard coupling mixture contained in a total volume of 0.15 mL 0.3 mg NacGlcHH2.BSA (Sigma Chemical Co., St. Louis, MO), 0.3 mg streptavidine (Pierce Chemical Co., Rockford, IL), 1.2 mg poly-L-lysine hydrochloride (Sigma Chemical Co.; molecular weight of 36700), 12.5 mM 1-ethyl-3(3-dimethylaminopropyl)carbodiimide (Pierce Chemical Co.), and 12.5 mM of N-hydroxysulfosuccinimide (Pierce Chemical Co.). Low-molecular-weight impurities in the proteins used in the coupling mixture were removed by Sephadex G-25 chromatography (Pharmacia Biotech, Uppsala, Sweden) before use. All reagents were dissolved in MilliQ water. The mixture was incubated at room temperature for 14-18 hours. Then 1 vol (0.15 mL) of 1 M of L-lysine was added, and the incubation was continued for another 4 hours. The coupling mixture was then stored at -20°C until purification. The conjugates were purified by high-performance liquid chromatography using a strong cation exchange column (Protein-Pak SP-8HR, 5x50mm; Water Associates, Milford, MA). After equilibration of the column with buffer A (0.02 M of sodium phosphate, pH 7.0), the coupling mixture was applied directly to the column. The column was washed with buffer A. The conjugate was eluted with a linear NaCl gradient (0-3 M) in buffer A. Fractions eluted between 1.45 M and 2.1 M of NaCl containing the conjugate were pooled, diluted 1:1 with water, concentrated four to six times using Ultrafree-CL filters UFC4LT-K25 (Waters Chromatography Division, Millipore Corp.), and stored at -20°C until use. The yield of the method was about 0.8 mg of purified conjugate per standard procedure.

Preparation of Adenovirus dl312

Propagation and purification of the replication-defective adenovirus mutant dl312 was performed as described previously with modifications (Graham et al., 1991). For large-scale preparation, the 293 cells were cultured to near confluency in four 750-mL Falcon flasks in IMDM, supplemented with 10% FCS. After aspiration of the culture medium, 3.5 mL of an adenovirus dl312 suspension, containing about 3.5×10^5 virus particles diluted in PBS, containing calcium and magnesium ions, was added to each flask. The cells were incubated for 1 hour at 37°C at 5% CO_2 atmosphere in a humidified incubator. The adenovirus suspension was then replaced with 30 mL of IMDM containing 2% FCS, and the incubation was continued until the full cytopathic effect was observed (about 60 hours). The adherent cells were detached using a rubber policeman. The cell suspension was centrifuged for 5 minutes at 175xg. The pellets from all four culture flasks were resuspended in a total of 5.0 mL of PBS and stored at -80°C. For virus purification, the cell suspension was transferred to an SS-34 Sorvall tube, containing 1g of washed, sterile glass beads (diameter 425-600 µm; Sigma Chemical, Co.). To lyse the cells, 0.6 mL of 5% sodium deoxycholate (Fluka, Buchs, Switzerland) was added. After vigorous mixing, the suspension was incubated on ice for 30 minutes. The thick cell lysate was vortexed for about 30-40 seconds to shear the chromosomal DNA, and the suspension was centrifuged twice (5000xg for 5 minutes at 4°C) to remove cell debris and glass beads. The virus was then purified by two rounds of centrifugation in a CsCl gradient, using two 5-mL quick-seal tubes each time. CsCl was removed by Sephadex G-25 chromatography in HEPES-buffered saline buffer (20 mM HEPES, pH 7.4; 150 mM NaCl; and 10% glycerol). The purified virus was concentrated using Ultrafree-CL filters UFC4 TMK25 (Millipore Corp.) and stored at -80°C.

Formation of DNA delivery complexes and incubation of cell cultures

The standard procedure for the DNA delivery complex formation was as follows: 2 µL conjugate (0.35-0.75 µg of protein) and 5 µL adenovirus mutant dl312 (diluted in HBS buffer; total 4×10^9 virus particles) were added successively to 193 µL of IMDM with vigorous mixing after each addition. After 30 minutes of incubation at room temperature, 50 µL DNA solution (48 µL IMDM plus 2 µL DNA diluted in HEPES-buffered saline) was added in small aliquots under continuous mixing, and the incubation was continued for an additional 30 minutes. After the addition of 28 µL of FCS and 222 µL IMDM supplemented with 10% FCS under continuous mixing, the delivery complex mixture was ready for use (final volume, 500 µL). For incubation with cells, after aspiration of the culture medium, 500 µL or 250 µL delivery complex mixture was added to each well of the 24- or 48-well Falcon culture plate. After 4 hours of incubation at 37°C, the delivery complex mixture was replaced by fresh IMDM supplemented with 10% FCS, followed by cell culture incubation. The chicken hepatoma cell line LMH was grown in IMDM, supplemented with 10% FCS. The cells were seeded at a density of 100,000-200,000 cells/mL, using 0.25 mL or 0.5 mL of cell suspension per well in 48- or 24-well Falcon culture plates (Becton Dickinson). Cell cultures were maintained routinely at 37°C and 5% CO_2 atmosphere in a humidified incubator.

Application of labeled oligonucleotides in Pekin ducks

Fluorescein-labeled oligonucleotides were injected into the foot vein of Pekin ducklings. After 2 hours the animals were sacrificed by injection of nembutal. The livers were removed and chilled in PBS. After digestion with collagenase the suspension was filtered through a fine wire-gauze mesh and centrifuged at 50xg for 2 minutes. The hepatocytes were resuspended in cold PBS and analyzed by FACSCAN.

Other methods

The plasmid pCMVluc contains the Photinus pyralis luciferase gene under the control of the cytomegalovirus promoter and enhancer. Plasmids were purified by two CsCl gradient centrifugations. The quantification of the activity of luciferase followed standard protocols, RLU were determined using a luminometer. The oligonucleotides (unmodified, phosphorothioate-modified, fluorescein-labeled) were purchased from Microsynth (Balgach, Switzerland).

4.6 SPECIAL NOTES

The coupling method to prepare protein conjugates is simple and reproducible. In addition to the known carbodiimide derivative 1-ethyl-3(3-dimethylaminopropyl) carbodiimide, N-hydroxysulfosuccinimide was used as a coupling enhancer. This allowed reduction of the concentration of 1-ethyl-3(3-dimethylaminopropyl) carbodiimide by more than 100 times and the use of both coupling reagents at low concentrations (12.5 mM). The best results were achieved when all of the protein constituents were coupled simultaneously. A 1:1:4 ratio of NAcGlcNH2.BSA:streptavidine:poly-L-lysine (wt/wt/wt) was found to be optimal for the coupling reaction. Streptavidine was included originally in the protein conjugate to form specific complexes with biotinylated adenovirus particles through the biotin-streptavidine bridge. Different from published data, in our system the adenovirus particles did not have to be biotinylated before complex formation with the conjugate. The complexes prepared with biotinylated or nonbiotinylated virus particles showed identical DNA delivery efficiencies. Despite this fact, delivery complexes containing a protein conjugate with streptavidine had an about 50% higher DNA delivery efficiency than those without streptavidine. The preparation of the delivery complexes had to be performed in serum-free medium. The addition of FCS before complex formation with DNA resulted in a decrease of delivery efficiency by more than 50%. Also, complex formation between adenoviral particles and protein conjugates was inhibited by FCS. Once the complete delivery complexes were formed, they were very stable in IMDM containing 10% FCS. After 12 hours of storage at 4°C, the delivery efficiency of the complexes remained unchanged ; after 24 hours of storage, the complexes still had a delivery efficiency of about 80%.

4.7 CONCLUDING REMARKS

Our data demonstrate the possibility to target the delivery of oligonucleotides to hepatocytes via receptor-mediated endocytosis using the ASGP-R. *In vitro* oligonucleotides can be delivered to hepatoma cells via this strategy, *in vivo* complexed unmodified oligonucleotides are taken up efficiently into the liver in contrast to non-complexed oligonucleotides which show only minimal uptake as determined by FACS analysis. The inclusion of replication-deficient adenoviruses in the delivery complexes is of central importance because of their endosomolytic activity. However, long-term cytopathic effects caused by replicational and/or transcriptional leakiness of the adenoviral mutant strain can cause serious problems. To circumvent the toxicity of adenoviral particles, adenoviruses from a distant species (human adenoviruses in Pekin duck cells) with very low infectivity were used in our experiments. Using the avian ASGP-R, specific transfer of the oligonucleotides to the liver can be achieved. Including the respective ligand for specific receptors of other tissues probably any desired tissue could be specifically transfected.

Experiments are now planned in which the inhibitory effect of complexed, unmodified oligonucleotides administered daily to DHBV-infected ducklings over a period of 7 days on DHBV replication will be tested. The aspect of possible side effects will be of special interest because of the known immunogenic properties of adenoviral particles and probably of the protein conjugates.

Furthermore, the delivery of plasmid DNA *in vivo* to the liver *via* this delivery strategy is of interest. Therefore, experiments with application of delivery complexes containing the reporter plasmid pCMVluc to the livers of ducklings *via* ASGP-R mediated endocytosis are underway.

4.8 ACKNOWLEDGEMENTS

This study was supported by a grant from the Deutsche Forschungsgemeinschaft (Of 14/4-1). W.-B. O. is the recipient of a Heisenberg-award from the Deutsche Forschungsgemeinschaft.

4.9 REFERENCES

Alt M, Caselmann WH. Liver-directed gene therapy: molecular tools and current preclinical and clinical studies. *J Hepatol* 1995;23:746-58

Bartholomew RM, Carmichael EP, Findeis MA, Wu CH, Wu GY. Targeted delivery of antisense DNA in woodchuck hepatitis virus-infected woodchucks. *J Viral Hepatitis* 1995;2:273-78

Bennett CF, Chiang M-Y, Chan H, Shoemaker JE, Mirabelli CK. Cationic lipids enhance cellular uptake and activity of phosphorothioate antisense oligonucleotides. *Mol Pharmacol* 1992;41:1023-33

Blum HE, Galun E, von Weizsäcker F, Wands JR. Inhibition of hepatitis B virus by antisense oligonucleotides. *Lancet* 1991;337:1230

Bongartz J-P, Aubertin A-M, Milhaud PG, Lebleu B. Improved biological activity of antisense oligonucleotides conjugated to a fusogenic peptide. *Nucleic Acids Res* 1994;22:4681-88

Boussif O, Lezoualc'h F, Zanta MA, Mergny MD, Scherman D, Demeneix B, Behr J-P. A versatile vector for gene and oligonucleotide transfer into cells in culture and in vivo: polyethylenimine. *Proc Natl Acad Sci USA* 1995;92:7297-301

Chang AGY, Wu GY. Gene therapy: applications to the treatment of gastrointestinal and liver diseases. *Gastroenterology* 1994;106:1076-84

Chemin I, Moradpour D, Wieland S, Offensperger WB, Walter E, Behr J-P, Blum HE. A linear polyethylenimine derivative mediates highly efficient gene and oligonucleotide transfer into primary hepatocytes. 1998 Submitted

Chowdhury NR, Wu CH, Wu GY, Yerneni PC, Bommineni VR, Chowdhury JR. Fate of DNA targeted to the liver by asialoglycoprotein receptor-mediated endocytosis in vivo. *J Biol Chem* 1993;268:11265-71

Compagnon B, Moradpour D, Alford DR, Larsen CE, Stevenson MJ, Mohr L, Wands JR, Nicolau C. Enhanced gene delivery and expression in human hepatocellular carcinoma cells by cationic immunoliposomes. *J Liposome Res* 1997;7:127-41

Condreay LD, Aldrich CE, Coates L, Mason WS, Wu TT. Efficient duck hepatitis B virus production by an avian liver tumor cell line. *J Virol* 1990;64:3249-58

Cotten M, Wagner E, Birnstiel ML. Receptor-mediated transport of DNA into eukaryotic cells. *Methods Enzymol* 1993a;217:618-45

Cotten M, Wagner E, Zatloukal K, Birnstiel ML. Chicken adenovirus (CELO) particles augment receptor-mediated DNA delivery to mammalian cells and yield exceptional levels of stable transformants. *J Virol* 1993b;67:3777-85

Cristiano RJ, Smith LC, Woo SL. Hepatic gene therapy: adenovirus enhancement of receptor-mediated gene delivery and expression in primary hepatocytes. *Proc Natl Acad Sci USA* 1993;90:2122-26

Curiel DT, Agarwal S, Wagner E, Cotten M. Adenovirus enhancement of transferrin-polylysine-mediated gene delivery. *Proc Natl Acad Sci USA* 1991;88:8850-54

Ferkol T, Perales JC, Eckman E, Kaetzel CS, Hanson RW, Davis PB. Gene transfer into the airway epithelium of animals by targeting the polymeric immunoglobulin receptor. *J Clin Invest* 1995;95:493-502

Ferkol T, Perales JC, Mularo F, Hanson RW. Receptor-mediated gene transfer into macrophages. *Proc Natl Acad Sci USA* 1996;93:101-05

Fiume L, Busi C, Mattioli A, Balboni PG, Barbanti-Brodano G. Hepatocyte targeting of adenine-9-beta-D-arabinofuranoside 5'-monophosphate (ara-AMP) coupled to lactosaminated albumin. *FEBS Lett* 1981;129:261-64

Gao X, Huang L. Cationic liposome-mediated gene transfer. *Gene Ther* 1995;2:710-22

Gewirtz AM, Stein CA, Glazer PM. Facilitating oligonucleotide delivery: helping antisense deliver on its promise. *Proc Natl Acad Sci USA* 1996;93:3161-63

Goodarzi G, Gross SC, Tewari A, Watabe A. Antisense oligonucleotides inhibit the expression of the gene for hepatitis B virus surface antigen. *J Gen Virol* 1990;71:3021-25

Graham FL, Prevec L. Manipulation of adenovirus vectors. *Methods Mol Biol* 1991;7:109-28

Hoofnagle JH, Di Bisceglie AM. The treatment of chronic viral hepatitis. *N Engl J Med* 1997;336:347-56

Jansen RW, Kruijt JK, Van Berkel TJ, Meijer DK. Coupling of the antiviral drug ara-AMP to lactosaminated albumin leads to specific uptake in rat and human hepatocytes. *Hepatology* 1993;18:146-52

Kamps JA, Morselt HW, Swart PJ, Meijer DK, Scherphof GL. Massive targeting of liposomes, surface-modified with anionized albumins, to hepatic endothelial cells. *Proc Natl Acad Sci USA* 1997;94:11681-85

Korba BE, Gerin JL. Antisense oligonucleotides are effective inhibitors of hepatitis B virus replication in vitro. *Antiviral Res* 1995;28:225-42

Lee WM. Hepatitis B virus infection. *N Engl J Med* 1997;337:1733-45

Ledley FD. Hepatic gene therapy. *Hepatology* 1993;18:1263-73

Ledley FD. Nonviral gene therapy: the promise of genes as pharmaceutical products. *Human Gene Ther* 1995;6:1129-44

Leonetti J-P, Machy P, Degols G, Lebleu B, Leserman L. Antibody-targeted liposomes containing oligodeoxyribonucleotides complementary to viral RNA selectively inhibit viral replication. *Proc Natl Acad Sci USA* 1990;87:2448-51

Letsinger RL, Zhang G, Sun DK, Ikeuchi T, Sarin PS. Cholesteryl-conjugated oligonucleotides: synthesis, properties, and activity as inhibitors of replication of human immunodeficiency virus in cell culture. *Proc Natl Acad Sci USA* 1989;86:6553-56

Lu X-M, Fischman AJ, Jyawook SL, Hendricks K, Tompkins RG, Yarmush ML. Antisense DNA delivery in vivo: liver targeting by receptor-mediated uptake. *J Nucl Med* 1994;35:269-75

Madon J, Blum HE. Receptor-mediated delivery of hepatitis B virus DNA and antisense oligonucleotides to avian liver cells. *Hepatology* 1996;24:474-81

Meijer DK, Molema G. Targeting of drugs to the liver. *Semin Liver Dis* 1995; 15: 202-56

Mellow TE, Halberg D, Drickamer K. Endocytosis of N-acetlyglucosamine-containing glycoproteins by rat fibroblasts expressing a single species of chicken liver glycoprotein receptor. *J Biol Chem* 1988;263:5468-73

Moradpour D, Compagnon B, Wilson BE, Nicolau C, Wands JR. Specific targeting of human hepatocellular carcinoma cells by immunoliposomes in vitro. *Hepatology* 1995;22:1527-37

Offensperger WB, Offensperger S, Walter E, Teubner K, Igloi G, Blum HE, Gerok W. In vivo inhibition of duck hepatitis B virus replication and gene expression by phosphorothioate-modified antisense oligodeoxynucleotides. *EMBO J* 1993;12:1257-62

Perales JC, Ferkol T, Beegen H, Ratnoff OD, Hanson RW. Gene transfer in vivo: sustained expression and regulation of genes introduced into the liver by receptor-targeted uptake. *Proc Natl Acad Sci USA* 1994;91:4086-90

Rensen PC, Van Dijk CM, Havenaar EC, Bijsterbosch MK, Kar Kruijt J, Van Berkel TJ. Selective liver targeting of antivirals by recombinant chylomicrons- a new therapeutic approach to hepatitis B. *Nat Med* 1995;1:221-25

Remy J-S, Kichler A, Mordvinov V, Schuber F, Behr J-P. Targeted gene transfer into hepatoma cells with lipopolyamine-condensed DNA particles presenting galactose ligands: a stage toward artificial viruses. *Proc Natl Acad Sci USA* 1995;92:1744-48

Soriano P, Dijkstra J, Legrand A, Spanjer H, Londos-Gagliardi D, Roerdink F, Scherphof G, et al. Targeted and nontargeted liposomes for in vivo transfer to rat liver cells of a plasmid containing the preproinsulin I gene. *Proc Natl Acad Sci USA* 1983;80:7128-31

Stankovics J, Andrews E, Wu G, Ledley FD. Overexpression of human methylmalonyl CoA mutase (MCM) in mouse liver after in vivo gene delivery using asialoglycoprotein complexes. *Am J Hum Genet* 1992;51:A177

Steer CJ. Receptor-mediated endocytosis: mechanisms, biologic function, and molecular properties. *Hepatology*, 1996;149-214. Zakim D, Boyer TD, eds. W. B. Saunders Company, Philadelphia

Szymkowski DE. Developing antisense oligonucleotides from the laboratory to clinical trials. *Drug Disc Tod* 1996;1:415-28

Torrani Cerenzia M, Fiume L, De Bernardi Venon W, Lavezzo B, Brunetto M, Ponzetto A, Di Stefano G, Busi C, Mattioli A, Gervasi G, Bonino F, Verme G. Adenine arabinoside monophosphate coupled to lactosaminated human albumin administered for 4 weeks in patients with chronic type B hepatitis decreased viremia without producing significant side effects. *Hepatology* 1996;23:657-61

Wagner E, Plank C, Zatloukal K, Cotten M, Birnstiel ML. Influenza virus hemagglutinine HA-2-N-terminal fusogenic peptides augment gene transfer by transferrin-polylysine complexes: toward a synthetic virus-like gene-transfer vehicle. *Proc Natl Acad Sci USA* 1992;89:7934-38

Wagner RW. Gene inhibition using antisense oligodeoxynucleotides. *Nature* 1994;372:333-35

Wang CY, Huang L. Plasmid DNA adsorbed to pH-sensitive liposomes efficiently transforms the target cells. *Biochem Biophys Res Commun* 1987;147:980-85

Wang S, Lee RJ, Cauchon G, Gorenstein DG, Low PS. Delivery of antisense oligodeoxyribonucleotides against the human epidermal growth factor receptor into cultured KB cells with liposomes conjugated to folate via polyethylene glycol. *Proc Natl Acad Sci USA* 1995;92:3318-22

Von Weizsäcker F, Wieland S, Köck J, Offensperger W-B, Offensperger S, Moradpour D, Blum HE. Gene therapy for chronic viral hepatitis: ribozymes, antisense oligonucleotides, and dominant negative mutants. *Hepatology* 1997;26:251-55

Wilson JM, Wu CH, Wu GY. Targeting genes: delivery and persistent expression of a foreign gene driven by mammalian regulatory elements in vivo. *J Biol Chem* 1989;264:16985-87

Wu G, Wu C. Receptor-mediated in vitro gene transformation by a soluble DNA carrier system. *J Biol Chem* 1987;262:4429-32

Wu GY, Wu CH. Receptor-mediated gene delivery and expression in vivo. *J Biol Chem* 1988;263:14621-24

Wu GY, Wilson JM, Shalaby F, Grossman M, Shafritz DA, Wu CH. Receptor-mediated gene delivery in vivo: partial correction of genetic analbumenia in Nagese rats. *J Biol Chem* 1991;266:14338-42

Wu G, Wu C. Specific inhibition of hepatitis B viral gene expression in vitro by targeted antisense oligonucleotides. *J Biol Chem* 1992;267:12436-39

Yao Z, Zhou Y, Feng X, Chen C, Guo J. In vivo inhibition of hepatitis B viral gene expression by antisense phosphorothioate oligodeoxynucleotides in athymic nude mice. *J Viral Hepatitis* 1996;3:19-22

5 How to Exclude Immunostimmulatory and Other Nonantisense Effects of Antisense Oligonucleotides

Arthur M. Krieg

Department of Internal Medicine
University of Iowa
Iowa City, IA, USA

5.1 INTRODUCTION

The use of antisense oligodeoxynucleotides for the sequence-specific modulation of gene expression is a tremendously versatile and exciting technology. However, like other polyanionic molecules, oligonucleotides have been found to have a variety of biologic activities that are independent of an antisense mechanism of action. These nonantisense activities, which can confound experimental interpretation, may be avoided through understanding the sequence motifs and responsible mechanisms, and through the careful use of appropriate control oligonucleotides.

5.2 OVERVIEW

A "dirty little secret" in the antisense field is that many of the biological effects of antisense oligonucleotides do not result from an antisense mechanism of action! In fact, it seems likely that the effects reported in many published "antisense" papers may result from nonantisense mechanisms that were not fully recognized or understood in the past. This chapter will review some of the nonantisense effects of oligonucleotides and discuss ways to design experiments so that these effects can be avoided or at least detected. By understanding the nonantisense activities of oligonucleotides, appropriate control oligonucleotides can be designed and tested to further enhance the experimenter's confidence in their ability to demonstrate an antisense effect.

Non-antisense effects of oligonucleotides

Non-sequence-specific effects.

Most of the oligonucleotide backbones commonly used in antisense are negatively charged, and the oligonucleotides are, therefore, polyanionic molecules. As such, they should be considered as drugs, and it should be appreciated that they can have biologic effects which are independent of hybridization to mRNA molecules through Watson-Crick base pairing. Many of these effects will depend on the oligonucleotide backbone. For example, the phosphorothioate oligonucleotide backbone has been reported to interact with a wide variety of different proteins (Stein, et al., 1993). Such protein binding by phosphorothioate oligonucleotide is presumably responsible for their sequence-independent concentration-dependent ability to induce the transcription factor, SP-1 (Perez, et al., 1994) and the JAK/STAT pathway (Too, 1998). Phosphorothioate oligonucleotide also interact with the extracellular matrix and bind both laminin and fibronectin, which can prevent cell adhesion (Khaled, et al., 1996). The occurrence of backbone-dependent effects such as these can generally be controlled for simply by using control oligonucleotides which have the same backbone.

Some of the nonantisense effects of oligonucleotides may only occur in the presence of other compounds. For example, low concentrations of phosphorothioate s may have no effects on cytokine secretion by themselves, but synergize with lipopolysaccharide (LPS) to increase tumor necrosis factor (TNF)-α secretion by fresh human peripheral blood mononuclear cells (PBMC) (Hartmann, et al., 1996). This effect appears to be another example of a polyanion activity of oligonucleotides since it is reversed by the polyanion heparin.

In many antisense experiments, proliferation assays are used as a readout for the inhibition of genes controlling cell proliferation. An important technical point for interpretation of experiments in which proliferation is measured by ^{3}H-thymidine incorporation assays concerns the possible effects of oligonucleotide degradation, especially when the antisense and control oligonucleotides differ in their thymidine content. Degradation of oligonucleotides containing thymidines cause the release of free thymidine, especially for thymidines present at the 3' end of an oligonucleotide (Matson and Krieg, 1992). Such thymidine released from the degraded oligonucleotides competes with ^{3}H-thymidine, thereby giving spuriously low proliferation values. Other assays for proliferation or assays for induction of ^{3}H-uridine incorporation can be used to avoid this artifact.

Sequence-specific effects.

oligonucleotides can have some sequence-selective non-antisense activities. For example, certain phosphorothioate oligonucleotides interact relatively specifically with the epidermal growth factor and vascular endothelial growth factor binding sites and can stimulate or antagonize receptor activation in a length-dependent

fashion (Rockwell, et al., 1997). Other types of sequence-specific non-antisense activities have also been recognized. Several sequence motifs have been reported which specifically inhibit various protein kinases, including repeated GGC sequences or a CGT(C)GA (Bergan, et al., 1994; Krieg, et al., 1997). In addition, an antisense oligonucleotide to the epidermal growth factor receptor, but not a sense oligonucleotide, was found to directly block the receptor kinase activity, but did not reduce the level of receptor expression (Coulson, 1996). oligonucleotides also can have sequence-specific effects to inhibit or induce cell proliferation. Degradation of the 3' ends of oligonucleotides releases the 5' monophosphate deoxyribonucleosides, which are toxic except for dCMP which can even neutralize the toxic effect of the other nucleosides (Vaerman, 1997).
Poly G motifs.

A sequence motif that can strongly inhibit cell proliferation is the so-called poly-G motif, which may consist of either four G's in a row or two G trimers in an oligonucleotide (Yaswen, et al., 1993; Burgess, et al., 1995). Lymphocytes can be induced to secrete IFN-γ by stimulation; but this is inhibited by G-rich oligonucleotides (Halpern and Pisetsky, 1995). Poly-G motifs in an oligonucleotide can interact with the antisense effects of the oligonucleotides in complex ways (Maltese, et al., 1995; Benimetskaya, et al., 1997). Although the anti-proliferative effects of poly-G motifs have only been described for oligonucleotides with a phosphorothioate backbone, poly-G motifs in phosphodiester oligonucleotides have also been reported to have non-antisense activities (Kimura, et al., 1994; Ramanathan, et al., 1994a; Ramanathan, et al., 1994b; Lee, et al., 1996). The ability of G-rich motifs to form intramolecular structures has been implicated in determining oligonucleotide nuclease resistance, cellular uptake and ability to block HIV infection (Bishop, et al., 1996). Furthermore, poly-G motifs in phosphorothioate oligonucleotides can bind to HIV-1 gp120 and prevent infection (Lederman, et al., 1996). For these reasons, investigators using antisense technology are best advised to avoid oligonucleotides containing either four Gs in a row or more than one region with three Gs in a row. However, even this caution may not be sufficient since some oligonucleotides containing G-rich regions such as four or five out of six consecutive bases G can also exert poly-G-like activities (Krieg, unpublished observation). Therefore, investigators using oligonucleotides that contain regions with significant G content should also test control oligonucleotides containing regions matched for their G content.

Sequence-specific immune activation by CpG motifs

A number of investigators working with "antisense" oligonucleotides have reported immune stimulatory effects. In experiments where the observed effects were anticipated or could be accounted for by an antisense mechanism of action, these have usually been concluded to result from an antisense activity. For example, we reported in 1989 that oligonucleotides antisense to certain endogenous retroviruses caused activation of lymphocyte proliferation (Krieg, et al., 1989). Although these results were highly reproducible, we subsequently discovered that the immune stimulation did not result from the supposed antisense mechanism of action, but rather was a consequence of the unrecognized coincidental presence of CpG dinucleotides in particular base contexts ("CpG motifs") in the antisense but not control oligonucleotides (Krieg, et al., 1995). In general, CpG dinucleotides are immune stimulatory if they are preceded by an A, G, or T. The immune stimulatory effect is most marked if the dinucleotide is also followed by one or more pyrimidines (Table 1). The stimulatory effects of CpG dinucleotides are lost or reduced if the CpG is preceded by a C or followed by a G (Yi, et al., 1996b). However, the effects of different CpG motifs are strongly dependent on the DNA backbone. Many different CpG motifs work with a phosphodiester backbone, but immune stimulation by phosphorothioate backbones is less predictable (Ballas, et al., 1996; Krieg, unpublished observation). Furthermore, the specific types of immune activation that are seen can vary depending on the backbone. For example, B cell proliferation is strongly induced by most phosphorothioate CpG motifs which are more effective than those with a phosphodiester backbone (Krieg, et al., 1995). However, induction of NK lytic activity is strong with a wide variety of phosphodiester CpG motifs, but is not seen with most phosphorothioate oligonucleotides unless they have the very strongest CpG motifs (Ballas, et al., 1996).

B cells and monocytes are activated by CpG DNA to generate reactive oxygen species, degrade IκB and translate NFκB into the nucleus, activate the mitogen-activated protein kinase signaling pathways, and induce transcription of multiple protooncogene and cytokine mRNAs (Krieg, et al., 1995; Yi, et al., 1996a; Yi, et al., 1996a; Yi, et al., 1998a,Yi, et al., 1998b; Klinman, et al., 1996; Stacey, et al., 1996; Sparwasser, et al., 1997). The CpG-induced activation pathway appears to be specifically sensitive to the effects of chloroquine and related antimalarial compounds, which completely block CpG-induced B cell and monocyte activation at low concentrations, but do not interfere with activation induced by endotoxin, anti-IgM, or ligation of CD40 (MacFarlane and Manzel, 1998; Yi, et al., 1998b).

CpG DNA induces immune activation which is characterized by B cell proliferation and IL-6 and immunoglobulin secretion; monocyte, macrophage, and dendritic cell production of IL-6, IL-12, TNF-α and type-1 interferons; and induction of natural killer cell lytic activity and IFN-γ secretion (Klinman, et al., 1996; Halpern, et al., 1996; Yamamoto, et al., 1992; Ballas, et al., 1996; Cowdery, et al., 1996).

Of interest, a recent study found that CpG dinucleotides are over-represented in published antisense oligonucleotide sequences compared to their expected frequency based on their occurrence in the human genome (Smetsers, et al., 1996). This raises the possibility that either such sequences may somehow mark mRNA regions susceptible to inhibition by antisense, or that biologic effects due to CpG dinucleotides may commonly be mistaken for antisense effects.

Aside from confounding the investigator wishing to pursue antisense experiments, immune activation by CpG motifs appears to play a role in normal immune defense mechanisms. It has long been appreciated that in vertebrate DNA, CpG dinucleotides occur less frequently than would be predicted by chance and are usually methylated on the cytosine at the 5 position (Bird, 1987). Furthermore, CpGs in vertebrate genomes are most commonly preceded by a C and/or followed by a G, which would make them nonimmune stimulatory. Thus, the presence of CpG motifs in prokaryotic DNA may enable it to be detected as foreign by the host immune system. It appears that the innate immune defense systems have evolved pattern recognition molecules which can detect molecular patterns that are present in microbes but not in our own cells. Because synthetic oligonucleotides are generally made with unmethylated cytosines, they can accidentally trigger this defense system if they contain one or more CpG motifs.

Avoidance of immune stimulatory effects of oligonucleotides.

The immune activation effects of CpG motifs can be avoided in antisense experiments through several approaches.
1. An obvious approach is simply to avoid any antisense oligo that contains a CpG. However, this may not be possible in all cases since it appears that not all regions of a target mRNA are accessible for antisense hybridization; and so the experimenter may be forced to target a region containing a CpG. If this is necessary, then the immune stimulatory effect can be avoided or reduced by substituting a 5-methyl cytosine during oligonucleotide synthesis (Krieg, et al., 1995; Table 1) or by bromo or iodo modifications to the 5 position of the cytosine ring (Boggs, et al., 1997).
2. We and others have discovered that the immune stimulatory effects of CpG DNA can be specifically blocked without interfering with other activation pathways by low concentrations of chloroquine and related antimalarial drugs (MacFarlane, et al., 1998; Yi, et al., 1998a). At the concentrations required to inhibit CpG immune stimulation (< 5 µM), these compounds do not block other pathways of cell activation such as those triggered by anti-IgM, LPS, PMA, or anti-CD40.
3. Another approach to avoid the CpG effects may be to use a cell type that does not respond to CpG DNA. To date, immune stimulatory effects have only been reported with B cells, NK cells, monocytes, macrophages, and dendritic cells. One group has reported that CpG motifs can costimulate T cell activation through the antigen receptor (Lipford, et al., 1997), but we have not observed T cell activation in our experimental systems. Very preliminary and incomplete studies have failed to detect cellular activation in studies of fibroblasts, endothelial cells, and epithelial cells. However, since no conclusion can be drawn from a negative experiment, it is probably prudent for the investigator using oligonucleotides with CpG motifs to test for possible stimulatory effects using control nonantisense oligonucleotides bearing the same CpG motif or oligonucleotides reported to have stimulatory effects in other systems. However, it is noteworthy that the immune effects of CpG motifs are somewhat species-specific; the CpG motifs that most strongly activate mouse cells have little or no

effect on human cells, although the CpG motifs that activate human cells do have a significant stimulatory effect on mouse cells (Krieg, et al., unpublished observation). Preliminary studies in cats indicate that they respond to distinct CpG motifs (Bruce Smith, Henry Baker and Arthur Krieg, unpublished data).

4. Use the lowest possible oligonucleotide concentration. This is good general advice for antisense experiments since most of the non-antisense effects of oligonucleotides are highly concentration-dependent. However, it can be difficult to find an oligonucleotide concentration that has an antisense effect without any risk of a CpG effect, since immune stimulation can be triggered by CpG oligonucleotide concentrations of 10 nM (Yi, et al., 1996a). Bennett, et al. reported that at concentrations of 50 nM or greater, phosphorothioate oligonucleotides not predicted to hybridize to VCAM-1 mRNA nevertheless reduced its expression (Bennett, et al., 1994).

5. Use an oligonucleotide backbone that does not support immune activation. The immune effects of CpG motifs are dependent on the DNA backbone and have only been observed with phosphodiester, phosphorothioate, or phosphorodithioate DNA backbones (Krieg, et al., 1996; Zhao, et al., 1996; Boggs, et al., 1997). As demonstrated in these studies, immune stimulation is minimal with oligonucleotides bearing methylphosphonate, phosphoramidate, p-ethoxy or, 2'-O-methyl backbones, or with a variety of specific base modifications.

Avoidance of other non-antisense effects of oligonucleotides

1. G-quartets (for example GAGGGG) have to be strictly avoided in antisense experiments. G-quartets in oligonucleotides lead to the formation of dimers and tetraplexes by Hoogsteen base pairs (in contrast to Watson-Crick base pairs), especially if G-quartets are located within two bases of either the 5' or 3' terminus of the oligonucleotide. For example, oligonucleotides containing G-quartets have been shown to block the expression of RelA protein (p65) by a non-antisense mechanism (Benimetskaya et al., 1997). The mechanism by which this occurs is not well defined.

2. Cleavage of the most 3'-terminal base by nuclease activity is associated with inhibition of cellular proliferation. This antiproliferative effect is due to the toxicity of the d-NMPs (5' monophosphate deoxyribonucleosides), the enzymatic hydrolysis products of the oligonucleotides. While d-CMP (2'deoxycytidine 5'- monophosphate) is not cytotoxic, all other d-NMPs are. Since only the Sp stereoisomer is nuclease resistant, this applies also to phosphorothioate oligonucleotides. Control oligonucleotides have to be designed adequately (Stein 1998; Vaerman et al., 1997).

Considerations in the design of control oligonucleotides

The control oligonucleotide that have been employed in antisense experiments may be divided into the following five classes:

1. Randomized oligonucleotides. In this type of control, each position in the oligonucleotide is randomized so that the control is actually a mixture of a very large number of different sequences. Randomized oligonucleotides can be used to demonstrate effects of an oligonucleotide backbone. However, this control should probably be considered obsolete since it is useless for determining the possible effects of any structural or sequence motifs that may be present in the antisense sequence.

2. Sense oligonucleotides. The standard kind of sense oligonucleotides will maintain general structural features of an oligonucleotide such as palindromes and other features that may form secondary structures but will not maintain compositional features such as G-rich sequences. A related type of control oligonucleotide has been termed the "inverse control" in which the antisense sequence is made by simply inverting the sequence of the antisense oligonucleotide so that the 3' end of the antisense oligonucleotide becomes the 5' end of the control sequence. This type of control will maintain base composition and should also be capable of forming similar types of secondary structures.

3. Scrambled oligonucleotides. This type of control, which probably should be considered obsolete, simply controls for base composition and backbone by scrambling the sequence of the antisense oligonucleotide.

4. Mismatched oligonucleotides. Usually, several different mismatched oligonucleotides are synthesized with two or more bases switched in the middle of the oligonucleotide to interfere with hybridization to the mRNA target while maintaining base composition of the antisense oligonucleotide. Depending on the selection of the bases to be switched, structural features and sequence motifs such as poly-G motifs and CpG motifs may or may not be maintained.

5. Mismatched target. This is the most rigorous type of control, and unfortunately is the least frequently performed. Cells treated with an antisense oligonucleotide are "rescued" by expression of the target mRNA with mutations to prevent hybridization with the antisense oligonucleotide.

Of course, it is essential that control oligonucleotides be made with the same backbone and other modifications as the antisense oligonucleotides. All oligonucleotides should be synthesized and purified using the same techniques and reagents. Careful use of appropriate control oligonucleotides is the single most important step in detecting non-antisense effects of oligonucleotides which could compromise the interpretation of an experiment.

Table 1: Effects of different CpG motifs on murine IL-6 production

ODN #	sequence 5'-3'	IL-6 (pg/ml)
Media		32.29
1916	TCCTGACGTTGAAGT	2847

1929	. . . . . GC	64.6
1936	. . . . . . ZG	1445
1937	. . Z . . . CG	2553
1917	. . . . . TCG	3124
1918	. . . . . GCG	2520
1919	. . . . . CCG	1079
1920	. . . . T . CG	1743.5
1921	. . . . A . CG	2558.5
1922	. . . . C . CG	2672.5
1923	. . . . . . CGA	1271
1924	. . . . . . CGC	1487
1925	. . . . . . CGG	209.7
1926	. . . . . . CG . A	1682
1927	. . . . . . CG . C	1742
1928	. . . . . . CG . G	1662
1930	. . . . . . CG . . . GG . G	1825.5
1931	. . . . . . CG . . CCTTC	3397
1935	. . . . . GCGGG	268.8
1938	. . . . AGCG	1675
1939	. . . A . . CG	5643
1940	. . . . . . CGGG	19
1626	GCA . . . CG . . . . GC .	1030

24 hr supernatants were collected from BALB/c spleen cells cultured at $5X10^6$/ml and assayed for IL-6 levels by ELISA essentially as described (Yi et al., 1996b). This is one of three experiments, which gave similar results.

Z = 5 methylcytosine

dots indicate identity to the original sequence 1916.

5.3 REFERENCES

Ballas ZK, Rasmussen WL, Krieg AM. Induction of natural killer activity in murine and human cells by CpG motifs in oligodeoxynucleotides and bacterial DNA. *J Immunol* 1996;157:1840-5.

Benimetskaya L, Berton M, Kolbanovsky A, Benimetsky S, Stein CA. Formation of a G-tetrad and higher order structures correlates with biological activity of the RelA (NF-κb p65) 'antisense' oligodeoxynucleotide. *Nucleic Acids Res* 1997;25:2648-56.

Bennett CF, Condon TP, Grimm S, Chan H, Chiang M-Y. Inhibition of endothelial cell adhesion molecule expression with antisense oligonucleotides. *J Immunol* 1994;152:3530-40.

Bergan R, Connell Y, Fahmy B, Kyle E, Neckers L. Aptameric inhibition of p210bcr-abl tyrosine kinase autophosphorylation by oligodeoxynucleotides of defined sequence and backbone structure. *Nucl Acids Res* 1994;22:2150.

Bird AP. CpG islands as gene markers in the vertebrate nucleus. *Trends Genet* 1987;3342-47.

Bishop JS, Guy-Caffee JK, Ojwang JO, Smith SR, Hogan ME, Cossum PA, Rando RF, Chaudhary N. Intramolecular G-quartet motifs confer nuclease resistance to a potent anti-HIV oligonucleotide. *J Biol Chem* 1996;271:5698-703.

Boggs RT, McGraw K, Condon T, Flournoy S, Villiet P, Bennett CF, Monia BP. Characterization and modulation of immune stimulation by modified oligonucleotides. *Antisense Nucl Acid Drug Dev* 1997;7:461-71.

Burgess TL, et al. The antiproliferative activity of c-myb and c-myc antisense oligonucleotides in smooth muscle cells is caused by a nonantisense mechanism. *Proc Natl Acad Sci USA* 1995;92:4051-55.

Coulson JM, Poyner DR, Chantry A, Irwin WJ, Akhtar S. A nonantisense sequence-selective effect of a phosphorothioate oligodeoxynucleotide directed against the epidermal growth factor receptor in A431 cells. *Mol Pharmacol* 1996;50:314-25

Cowdery JS, Chace JH, Krieg AM. Bacterial DNA induces in vivo interferon-γ production by NK cells and increases sensitivity to endotoxin. *J Immunol* 1996;156:4570-5.

Halpern HD, Pisetsky DS. In vitro inhibition of murine IFN-γ production by phosphorothioate deoxyguanosine oligomers. *Immunopharmacology* 1995;29:47-52.

Halpern MD, Kurlander RJ, Pisetsky DS. Bacterial DNA induces murine interferon-γ production by stimulation of interleukin-12 and tumor necrosis factor-α. *Cell Immunol* 1996;167:72-8.

Hartmann G, Krug A, Waller-Fontaine K, Endres S. Oligodeoxynucleotides enhance lipopolysaccharide-stimulated synthesis of tumor necrosis factor: dependence on phosphorothioate modification and reversal by heparin. *Mol Med* 1996;2:429-38.

Khaled Z, Benimetskaya L, Zeltser R, Khan T, Sharma HW, Narayanan R, Stein CA. Multiple mechanisms may contribute to the cellular anti-adhesive effects of phosphorothioate oligodeoxynucleotides. *Nucleic Acids Res* 1996;24:737-45.

Kimura Y, Sonehara K, Kuramoto E, Makino T, Yamamoto S, Yamamoto T, Kataoka T, Tokunaga T. Binding of oligoguanylate to scavenger receptors is required for oligonucleotides to augment NK cell activity and induce IFN. *J Biochem* 1994;116:991-4.

Klinman D, Yi A-K, Beaucage SL, Conover J, Krieg AM. CpG motifs expressed by bacterial DNA rapidly induce lymphocytes to secrete IL-6, IL-12 and IFN. *Proc Natl Acad Sci USA* 1996;93:2879-83.

Krieg AM, Gause WC, Gourley MF, Steinberg AD. A role for endogenous retroviral sequences in the regulation of lymphocyte activation. *J Immunol* 1989;143:2448-51.

Krieg AK, Yi A-K, Matson S, Waldschmidt TJ, Bishop GA, Teasdale R, Koretzky G, Klinman D. CpG motifs in bacterial DNA trigger direct B-cell activation. *Nature* 1995;374:546-9.

Krieg AM, Matson S, Herrera C, Fisher E. Oligodeoxynucleotide modifications determine the magnitude of immune stimulation by CpG motifs. *Antisense Res Dev* 1996;6:133-9.

Krieg AM, Matson S, Cheng K, Fisher E, Koretzky GA, Koland JG. Identification of an oligodeoxynucleotide sequence motif that specifically inhibits phosphorylation by protein tyrosine kinases. *Antisense Nucl Acid Drug Dev* 1997;7:115-123.

Lederman S, Sullivan G, Benimetskaya L, Lowy I, Land K, Khaled Z, Cleary AM, Yakubov L, Stein CA. Polydeoxyguanine motifs in a 12-mer phosphorothioate oligodeoxynucletide augment binding to the v3 loop of HIV-1 gp120 and potency of HIV-1 inhibition independency of G-tetrad formation. *Antisense Nucl Drug Dev* 1996;6:281-9.

Lee PP, Ramanathan M, Hunt CA, Garovoy MR. An oligonucleotide blocks interferon-γ signal transduction. *Transplantation* 1996;62:1297-1301.

Lipford GB, Bauer M, Blank C, Reiter R, Wagner H, Heeg K. CpG-containing synthetic oligonucleotides promote B and cytotoxic T cell responses to protein antigen: a new class of vaccine adjuvants. *Eur J Immunol* 1997;27:2340-4.

MacFarlane DE, Manzel L. Antagonism of immunostimulatory CpG-oligodeoxynucletides by quinacrine, chloroquine, and structurally related compounds. *J Immunol* 1998;160:1122-31.

Maltese JY, Sharma HW, Vassilev L, Narayanan R. Sequence context of antisense RelA/NF-κB phosphorothioates determines specificity. *Nucleic Acids Res* 1995;23:1146-51.

Matson S, Krieg AM. Nonspecific suppression of 3H-thymidine incorporation by "control" oligonucleotides. *Antisense Res Dev* 1992;2:325-330.

Perez JR, Li Y, Stein CA, Majumder S, Van Oorschot A, Narayanan R. Sequence-independent induction of Sp1 transcription factor activity by phosphorothioate oligodeoxynucleotides. *Proc Natl Acad Sci USA* 1994;91:5957-61.

Ramanathan M, Lantz M, MacGregor RD, Garovoy MR, Hunt CA. Characterization of the oligodeoxynucleotide-mediated inhibition of interferon-γ-induced major histocompatibility complex class I and intercellular adhesion molecule-1. *J Biol Chem* 1994a ;269:24564-74.

Ramanathan M, Lantz M, MacGregor RD, Huey B, Tam S, Ki Y, Garovoy MR., Hunt CA. Inhibition of interferon-γ-induced major histocompatibility complex class I expression by certain oligodeoxynucleotides. *Transplantation* 1994b ;57:612-5.

Rockwell P, O'Conner WJ, King K, Goldstein NI, Zhang LM, Stein CA. Cell-surface perturbations of the epidermal growth factor and vascular endothelial growth factor receptors by phosphorothioate oligodeoxynucleotides. *Proc Natl Acad Sci USA* 1997;94:6523-8.

Smetsers TFCM, Boezeman JBM, Mensink EJBM. Bias in nucleotide composition of antisense oligonucleotides. *Antisense Nucl Acid Drug Dev* 1996;6:63-7.

Sparwasser T, Miethe T, Lipford G, Erdmann A, Hacker H, Heeg K, Wagner H. Macrophages sense pathogens via DNA motifs: induction of tumor necrosis factor-α-mediated shock. *Eur J Immunol* 1997;27:1671-9.

Stacey KJ, Sweet MJ, Hume DA. Macrophages ingest and are activated by bacterial DNA. *J Immunol* 1996;157:2116-22.

Stein CA (1998) How to design an antisense oligodeoxynucleotide experiment: a consensus approach. *Antisense Nucleic Acid Drug Dev* 8:129-132.

Stein CA, Tonkinson JL, Zhang L-M, Yakubov L, Gervasoni J, Taub R, Rotenberg SA. Dynamics of the internalization of phosphodiester oligodeoxynucleotides in HL60 cells. *Biochem* 1993;32:4855-61.

Too CKL. Rapid induction of Jak2 and Sp1 in T cells by phosphorothioate oligonucleotides. *Antisense Nucl Acid Drug Dev* 1998;8:87-94.

Vaerman JL, Moureau P, Deldime F, Lewalle P, Lammineur C, Morschhauser F, Martiat P. Antisense oligodeoxyribonucleotides suppress hematologic cell growth through stepwise release of deoxyribonucleotides. *Blood* 1997;90:331-9.

Yamamoto S, Yamamoto T, Kataoka T, Kuramoto E, Yano O, Tokunaga T. Unique palindromic sequences in synthetic oligonucleotides are required to induce INF and augment INF-mediated natural killer activity. *J Immunol* 1992;148:4072-6.

Yaswen P, Stampfer MR, Ghosh K, Cohen JS. Effects of sequence of thioated oligonucleotides on cultured human mammary epithelial cells. *Antisense Res Dev* 1993;3:67-77.

Yi A-K, Hornbeck P, Lafrenz DE, Krieg AM. CpG DNA rescue of murine B lymphoma cells from anti-IgM induced growth arrest and programmed cell death is associated with increased expression of c-myc and bcl-XL. *J Immunol* 1996a ;157:4918-25.

Yi A-K, Klinman DM, Martin TL, Matson S, Krieg AM. Rapid immune activation by CpG motifs in bacterial DNA: Systemic induction of IL-6 transcription through an antioxidant-sensitive pathway. *J Immunol* 1996b;157:5394-402.

Yi A-K, Tuetken R, Redford T, Kirsch J, Krieg AM. CpG motifs in bacterial DNA activates B cells and monocytes through the pH-dependent generation of reactive oxygen species. *J Immunol* 1998a;in press.

Yi A-K, Chang M, Peckham DW, Krieg AM, Ashman RF. CpG oligodeoxyribonucleotides rescue mature spleen B cells from spontaneous apoptosis and promote cell cycle entry. *J Immunol* 1998b;in press.

Zhao Q, Temsamani J, Iadarola PL, Jiang Z, Agrawal S. Effect of different chemically modified oligodeoxynucleotides on immune stimulation. *Biochem Pharmac* 1996;51:173-82.

6 Testing Antisense Oligonucleotides in Controlled Cell Culture Assays

W. Michael Flanagan

Department of Cell Biology
Gilead Sciences

6.1 INTRODUCTION

Antisense oligonucleotides (oligonucleotides) are short synthetic nucleic acids that contain complementary base sequences to a targeted RNA. Hybridization of the RNA with the antisense oligonucleotide interferes with RNA function and ultimately blocks protein expression. Therefore, any gene for which the partial sequence is known can be targeted by an antisense oligonucleotide.

While the concept of antisense inhibition is elegantly simple, the development of antisense-based reagents to elucidate and validate the role of a particular gene in a complex biological system has been difficult. However, recent technological advances and an understanding of the limitations of antisense-based compounds has resulted in the evolution of a robust laboratory tool (Flanagan et al., 1997; Wagner et al., 1997; Matteucci et al., 1996; Wagner, 1995a; Wagner, 1995b; Dean et al., 1996; Stein et al., 1995).

In this chapter, I will discuss our experience using antisense-based inhibitors including the selection of antisense sequences, modifications of antisense oligonucleotides, oligonucleotide delivery methods, and guidelines for the analysis of antisense results. The goal of this review is to educate the antisense neophyte so that common experimental pitfalls can be avoided and informative data can be obtained using antisense oligonucleotides.

Selection of antisense oligonucleotides

Although several methods have recently been proposed to aid the identification of antisense oligonucleotides, as discussed in a previous chapter, in most laboratories the selection of active antisense oligonucleotides is still empirical (Ho et al., 1998; Patzel et al., 1998). Traditionally, researchers have screened two or three antisense oligonucleotides targeting the translational start site of a mRNA, based on the largely incorrect assumption that antisense oligonucleotides function by sterically blocking ribosome entry or progression along the mRNA. Most conventional antisense oligonucleotides (unmodified and C-5 propyne modified phosphorothioate oligonucleotides) function through a RNase H mechanism of action (Moulds et al., 1995, Wagner et al., 1993). RNase H is a endogenous cellular enzyme that selectively hydrolyzes the RNA portion of the oligonucleotide/RNA heteroduplex and is required for potent antisense specific inhibition. Using RNase H competent antisense oligonucleotides, any region of the RNA can be targeted including the 5' and 3' untranslated regions, introns, the coding region, and the start of translation; however, as discussed below, several factors can influence the success rate of identifying active antisense oligonucleotides.

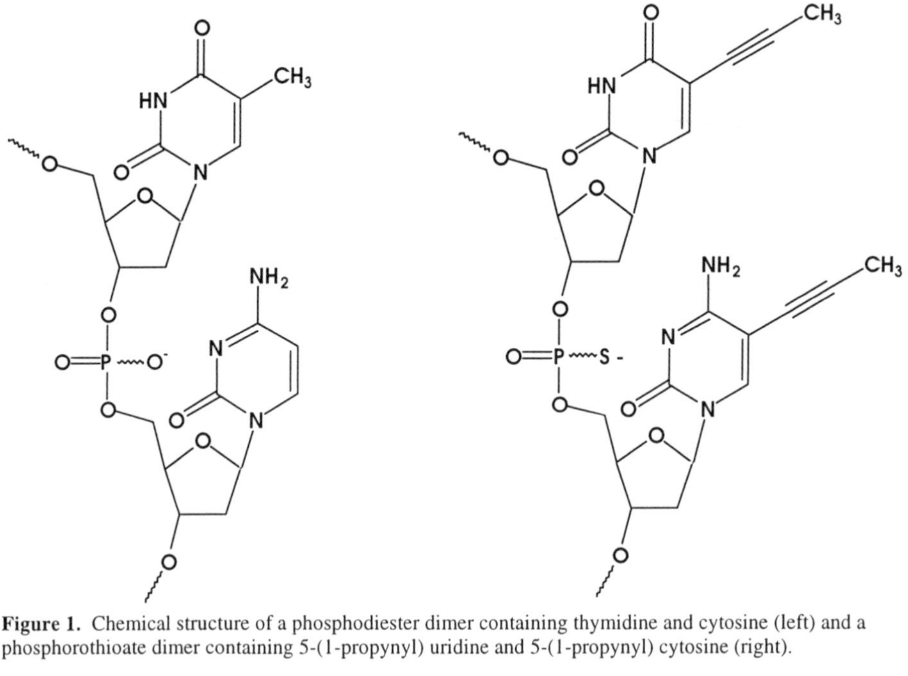

Figure 1. Chemical structure of a phosphodiester dimer containing thymidine and cytosine (left) and a phosphorothioate dimer containing 5-(1-propynyl) uridine and 5-(1-propynyl) cytosine (right).

Oligonucleotide modifications

Chemical stability and high affinity for the target sequence are two characteristics that are required for an antisense oligonucleotide to be successful. The earliest antisense oligonucleotides tested *in vitro* contained phosphodiester backbones (Zamecnik et al., 1978). Although several publications reported oligonucleotide-dependent antisense effects using phosphodiester oligonucleotides, it is unlikely these observed biological effects were due to true antisense mechanism of action since it was subsequently shown that phosphodiester containing oligonucleotide are rapidly degraded (<20 minutes) both in tissue culture media and in the cell (Fisher et al., 1993). This problem was overcome by replacing one of the non-bridging oxygen atoms of the phosphodiester backbone with a sulfur atom (Figure 1) which created a nuclease resistant phosphorothioate backbone that retained the ability to recruit RNase H (Shaw et al., 1991). Currently, phosphorothioate modified oligonucleotides are the most widely used and commercially available oligonucleotides for antisense research. However, one drawback to phosphorothioate modified oligonucleotides is that they have decreased binding affinity for their RNA target compared to their phosphodiester containing predecessors. Because of the decrease affinity of phosphorothioate oligonucleotides, researchers usually need to screen 20-50 different 18-21 nucleotide long oligonucleotides to identify an antisense oligonucleotide that demonstrates specific and potent antisense activity (Monia et al., 1996a).

C-5 Propynyl pyrimidines

These shortcomings prompted us to search for oligonucleotide modifications that would increase the affinity of oligonucleotides for RNA. Methyl substitutions at the C-5 position of pyrimidines were known to increase the affinity of oligonucleotide for RNA. A systematic synthesis and evaluation of a large number of modifications at the C-5 position resulted in the identification of C-5 propyne derivatives of uracil and cytosine (Froehler et al., 1992) (Figure 1). The incorporation of C-5 propyne modifications into phosphorothioate oligonucleotides dramatically augmented the affinity of oligonucleotides for RNA based on measurements of thermal melting temperature, and resulted in greater than 10-fold more potent antisense effects *in vitro* as compared to regular phosphorothioate oligonucleotides (Wagner et al., 1993). In addition, several potent and specific propyne-modified antisense oligonucleotides can be readily identified from screening as few as six antisense oligonucleotides as long as the oligonucleotide sequences are selected based on the folllowing criteria: (1) the oligonucleotide is 50-80% pyrimidine to fully utilize the C-5 propynyl pyrimidines; and (2) the oligonucleotide is 15-nt in length which is long enough to provide stability upon binding and uniqueness in the human genome. We routinely use C-5 propyne modified phosphorothioate antisense oligonucleotides for all our *in vitro* experiments (Flanagan et al., 1997; Flanagan et al., 1996a; Flanagan et al., 1996b; Lewis et al., 1996; St. Croix et al., 1996; Coats et al., 1996). The C-5 propyne phosphoramidites are available from Glen Research, Sterling, VA.

6.2 EVALUATING ANTISENSE ACTIVITY

Phosphorothioate oligonucleotides can have a wide variety of sequence-dependent, non-antisense effects (Stein, 1995; Stein, 1996). For instance, oligonucleotides containing a guanosine quartet have been shown to have antiproliferative activity (Burgess et al., 1995). Phosphorothioate oligonucleotides can also interact with extracellular receptors and profoundly affect cellular metabolism and differentiation (Benimetskaya et al., 1997; Guvakova et al., 1995), or block virion binding and fusion (Azad et al., 1993). Futhermore, as discussed in a previous chapter, oligonucleotides containing CpG dimers can activate B cells and natural killer cells which could result in immune stimulation and antitumor activity *in vivo* (Krieg et al., 1995). Because of the wide variety of potential nonspecific effects, the evaluation of antisense oligonucleotide activity should never be based solely on their phenotypic or biological effects *in vitro* or *in vivo* rather direct measurement of the targeted RNA and protein should be conducted as discussed below (Monia, 1997; Wagner, 1994; Cowsert, 1997).

Experimental guidelines

Many of these nonspecific effects of phosphorothioate oligonucleotides can be identified and avoided by including rigorous controls in antisense studies. The most convincing and easily published antisense experiments include the following experimental controls (Dean et al., 1996; Stein et al., 1995).
1. Direct measurement of the effect of the oligonucleotide on expression of the targeted RNA or protein levels as compared to its effect on an internal control RNA or protein.
2. Demonstration that antisense inhibition of the targeted RNA or protein is dose-dependent.
3. Demonstration of a rank order potency of several unrelated antisense oligonucleotides. These data suggest that inhibition of the targeted RNA or protein is sequence-dependent rather than a general non-specific effect of phosphorothioate oligonucleotides.
4. Liberal use of control oligonucleotides including scramble and mismatch controls that retain the same overall base composition as the antisense oligonucleotide but have a different sequence. Futhermore, correlation of antisense activity of the oligonucleotide and the effect of increasing number of incorporated mismatches into the oligonucleotide provides a compelling argument that the oligonucleotide is working through an antisense mechanism of action (Monia et al., 1996b).
5. One of the most rigorous experimental controls is genetic complementation of the antisense effect. For example, Coats et al. showed that antisense inhibition of p27kip1, a cyclin dependent kinase inhibitor, allowed cells to continue to proliferate in the absence of mitogens (Coats et al., 1996). To demonstrate that the cell cycle effect was due to loss of p27kip1, they complemented the antisense-treated cells with a p27kip1 expression plasmid that could not be inhibited by the antisense oligonucleotides. Mutations were constructed that, by exploiting the degeneracy of the genetic code, altered the antisense binding site yet encoded for wild-type p27kip1. Enforced expression of the sequence altered p27kip1 mRNA restored mitogen responsiveness. These results rigorously showed that the inability of p27kip1 antisense-treated cells to exit the cell cycle after mitogen starvation was specifically caused by the loss of p27kip1 expression.
When these experimental guidelines are followed, antisense oligonucleotides can be very powerful research tools and can be confidently used as highly specific and

potent inhibitors to elucidate gene function and validate targets in a wide variety of biological systems.

6.3 ON DELIVERY TO TISSUE CULTURE CELLS

Phosphorothioate antisense oligonucleotides are large (molecular weight of about 6000), highly charged molecules that are poorly internalized by most tissue culture cells. Furthermore, if they are internalized, it is not clear that they can escape from endosomes and lysosomes and accumulate in the nucleus. The nucleus of the cell is the major site of action for RNase H competent antisense oligonucleotides (Moulds et al., 1995). Despite the on going controversy regarding whether oligonucleotide can enter tissue culture cells unaided, most seasoned antisense researchers use a variety of techniques to introduce antisense oligonucleotides into cells including microinjection (Wagner et al., 1993), cationic liposomes (Lewis et al., 1996; Bennett et al., 1992), and electroporation (Bergan et al., 1993)

Microinjection of antisense oligonucleotides

Microinjection of oligonucleotides into tissue culture cells is a convenient way to screen and evaluate antisense oligonucleotides *in vitro* (reviewed in Flanagan et al., 1997). A control expression plasmid is coinjected with the target plasmid directly into the nucleus of the cell along with the oligonucleotide. Expression from the control plasmid is monitored for non-specific effects of the antisense oligonucleotides being evaluted and serves as a marker for injected cells. Expression from the target plasmid is examined in the presence and absence of the antisense oligonucleotides being tested. The most potent antisense oligonucleotides completely block expression from the targeted plasmid while having no effect on expression of the reporter plasmid. In our lab, we routinely use the microinjection assay (Wagner et al., 1993; Hanvey et al., 1992; Graessmann et al., 1994) to assess new chemical modifications of antisense oligonucleotides and to precisely determine the rank order potency of different modified antisense oligonucleotides.

Protocol for microinjection of antisense oligonucleotides into adherent cells

Materials required

Microinjection is performed on an inverted microscope (Zeiss Axiovert 10) equipped with a 32X Achrostigmat objective and 10X eyepieces and mounted on a vibration-free floating table. Glass capillary pipettes are prepared using a needle puller (Sutter Instruments, Novato, CA). The needle is attached to a micromanipulator (Narishige MO-302) and controlled by a pneumatic controller (Nikon PLI-1888).

Tissue culture cells

Cells are grown on 25 mm #1 glass circular coverslips in Dulbecco's modified Eagles media (DMEM) and 10% fetal bovine serum (FBS) in humidified incubators at 37° C with 5% CO_2 Immediately before injection, the cells are mounted on a temperature-controlled chamber (34°C) and the media is replaced with DMEM + 10% FBS containing 50 mM HEPES (pH 7.3). The HEPES buffered media maintains the pH optimum during the microinjection procedure. Black waterproof ink marks are made on the bottom of the glass slide to provide landmarks on the slide for a new microinjection line. The cells are microinjected in a straight line across the slide to simplify sample identification and later analysis.

Sample preparation

Plasmids are prepared by cesium chloride banding, Qiagen (Qiagen, Chatsworth, CA) maxiprep kits, or Promega (Promega, Madison, WI) wizard minpreps all express well following microinjection. Expression plasmids that use the cytomegalovirus (CMV) promoter/enhancer to promote expression work the best in our experience. The total number of plasmids injected per cell should be between 100-1000 copies. The plasmid copy number should be titrated to obtain uniform expression in all the cells injected.

1X Microinjection buffer

50 mM HEPES (pH 7.3)
90 mM KCl

Oligonucleotide preparation

C-5 propyne oligonucleotides are synthesized as previously described (Flanagan et al., 1997). The C-5 propyne oligonucleotides are resuspended to 500 μM in sterile water. Oligonucleotides can be quantitated based on a calculated extinction coefficient at A_{260nm} using the following information

T base =	$8.8 \times 10^3 \, cm^{-1}M^{-1}$
C base =	$7.4 \times 10^3 \, cm^{-1}M^{-1}$
pdU base =	$3.2 \times 10^3 \, cm^{-1}M^{-1}$
pdC base =	$5.0 \times 10^3 \, cm^{-1}M^{-1}$
A base =	$15.2 \times 10^3 \, cm^{-1}M^{-1}$
G base =	$11.8 \times 10^3 \, cm^{-1}M^{-1}$

Estimation of intracellular oligonucleotide concentration

Active C-5 propyne modified phosphorothioate oligonucleotides show complete suppression of gene expression at 10 μM needle concentration (for 4 to 24 h post-injection); approximate injection volumes are 10–20 femtoliters (fl, 10^{-15}) for nuclear injection (~1/20 the cell volume) which translates to about 0.5 μM intracellular oligonucleotide concentration (Wagner et al., 1993).

Microinjection procedure

1. The temperature-controlled chamber with the cells grown on a glass coverslip is mounted onto the microscope stage and the cells are brought into focus. The glass needle is lowered and brought into focus just above the cell to be injected.
2. The needle is then lowered until the tip enters the cell. Just before the tip fully enters the cell a depression on the cell surface is seen. The depression appears as a small white spot on the cell surface.
3. The plasmids (control plasmid and target plasmid) and oligonucleotides are injected into the cell by a low pressure injection pump. When injection of the material occurs, a small puff or slight enlargement of the cell is seen. If no puff is seen upon injection, the needle may be clogged and should be replaced with a new needle and the procedure repeated.
4. Following injection, the needle is gently raised up and out of the cell and positioned above the next cell to be injected. With some practice, approximately 100-200 cells can be injected in 15 minutes.
5. Upon completion of all the injections for the glass slide, the slide is removed from the mount and placed in a humidified incubator with 5% CO_2 at 37°C for 4.5 to 48 hours. The cells are then fixed and immunostained for detection of the expressed proteins (see the immunofluorescent staining protocol below) (Flanagan et al., 1997; Wagner et al., 1993; Hanvey et al., 1992).

Immunofluorescent staining

Instrumentation required

Microscopes:
1. Inverted fluorescent microscope with 50W or 100W mercury illumination
        (we use either a Zeiss Axiovert 10 or a Nikon Diaphot)
2. Filters: Omega Optical (Brattleboro, VT)
        Fluorescein: Excit.: 480DF30, Dichr.: 505DRLP, Emiss: 535DF45
        Texas Red: Excit.: 560DF40, Dichr.: 595DRLP, Emiss: 635DF60
3. Lenses:
        Nikon: 60x Phase 4 oil lens for fluorescence, NA1.3
        Zeiss: 63x NA1.3 lens for fluorescence.

Immunostaining protocol

1. Rinse cells at room temperature with Tris-buffered saline (TBS: 10 mM Tris Cl, pH 7.3, 0.1 M NaCl).
2. Fix 15 min in 3.7% formaldehyde solution prepared fresh by diluting 37% formaldehyde stock solution (Sigma, St. Louis, MO) into Hepes-buffered saline (HS: 10 mM Hepes, pH 7.3, 0.1 M NaCl).
3. Rinse the cells twice with TBS without letting the slides dry out between washes.
4. Soak the cells in TBS + 50 mM glycine (10 min) to quench any residual formaldehyde.
5. For intracellular antigens: permeabilize the cells 5 min (up to 10 min) in 1% Nonidet P-40 (NP40; Sigma) in TBS. Some antigens are better unmasked and stained when cells are permeabilized at -20°C in cold methanol.
6. Wash the cells twice with antibody wash (Ab wash; TBS containing 0.2% NP40). For surface antigens, use TBS with no detergents since detergents may strip the antigen of interest from the membrane.
7. Block nonspecific sites in the cells with filter-sterilized 5% BSA in Ab wash containing 0.1% NaN_3 (from 15 min to overnight). Some researchers block with goat serum or instant milk or combinations thereof to reduce background.
8. Primary antibody: Dilute Ab in the 5% BSA solution described above, mix by inversion, and microfuge 5 min, full speed at room temperature. Incubate Ab on the slide for a minimum of 45 minutes. To minimize Ab use, place the coverslip on a piece of parafilm. Routinely, we use the primary antibody at a final concentration of 0.1 µg/ml although it is important to consult the suppliers recommendations before using any antibody. For a 25 mm coverslip, 90–100 µl of solution is sufficient. Put a wet Kimwipe off to the side to prevent cells from drying out and cover (a 35 mm petri dish top works well).
9. Wash three times with Ab wash and allow 5 min for each wash. Longer wash times (10-15 min) usually result in lower background interference. For surface antigens, wash in TBS without detergents.
10. Fluorescently conjugated secondary antibody. Repeat step 8 but on a fresh sheet of parafilm. We also use the secondary antibody at 0.1 µg/ml. We purchase all our secondary antibodies from Jackson ImmunoResearch Laboratories, Inc. (West Grove, PA). When using a fluorescent secondary Ab, 30 min incubation is usually sufficient.
11. Wash as described in #9 above.
12. For some expressed protein targets, a tertiary antibody is required to generate sufficient signal to detect protein expression. If this is the case, repeat steps 10 and 11 with an appropriate fluorescently conjugated tertiary antibody.
13. Rinse coverslip in distilled water, absorb excess water with edge of Kimwipe, and mount with the stained cells face down on microscope slidewith Fluoromount G (Southern Biotechnology Association, Inc., Birmingham, AL). Aspirate excess Fluoromount G and seal the edge of the glass slide to the microscope slide with clear nail polish.

Comments

The microinjection assay is a reliable method for quickly identifying and evaluating active antisense oligonucleotides. Occasionally technical problems have occurred in which active oligonucleotides have failed to demonstrate antisense activity in this assay. To avoid these pitfalls, several precautions can be taken such as using expression plasmids, microinjection buffer, and oligonucleotides that are freshly made on a monthly basis and stored at –20°C in small aliquots. In addition, tissue culture cells should be used within about 10 passages following a thaw to obtain consistent antisense results.

6.4 CATIONIC LIPID MEDIATED OLIGONUCLEOTIDE TRANSFECTION

Introducing DNA into cells with cationic lipids can be a powerful tool for examining the roles of genes in biological systems. Most lipid mediated transfection reagents have three major drawbacks:poor transfection efficiency, cytotoxicity, and they must be used in serum-free growth media. GS 3815 cytofectin alleviates all of these problems. GS 3815 cytofectin (Glen Research Inc., Sterling, VA) delivers oligonucleotides and DNA efficiently to a broad spectrum of cell lines in the presence of serum containing growth media with little or no cytotoxicity. GS 3815 cytofectin is a formulation of a cationic lipid shown in Figure 2, with the zwitterion L-a Dioleoyl Phosphatidylethanolamine (DOPE, Avanti Polar Lipids, Inc., Alabaster, AL). GS3815 cytofectin can deliver oligonucleotides as well as plasmids.

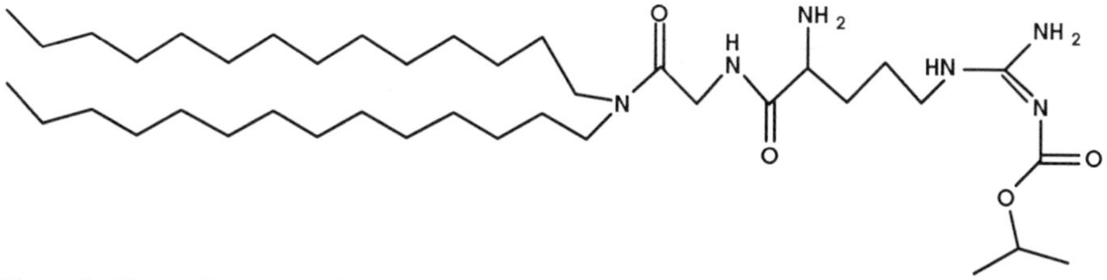

Figure 2. Chemical structure of GS3815 cytofectin.

The following is a protocol for using GS3815 cytofectin to transfect antisense phosphorothioate oligonucleotides into adherent cells grown on 100 mm tissue culture dish (Falcon #3003, Becton Dickinson Labware; Plymouth, England). This protocol is a modification of previous protocols [Flanagan et al., 1997; Lewis et al., 1996).

Stability and storage conditions

1. Store lipid at 4°C; mix well before use. Lipid is stable for at least three months at 4°C. For long term storage keep at –70°C. GS3815 cytofectin is a liquid formulation of 2 moles of GS3815 cationic lipid and one mole of DOPE and is commericially available as a 1 mg/ml stock in 1ml aliqouts. The molecular weight (mw) of GS3815 is 779 g/mol and the mw of DOPE is 744 g/mol.
2. Store oligonucleotides at -20°C. If you are going to be using the oligonucleotide frequently, just tore it on your bench at room temperature to avoid repeated freeze/thaw cycles.

Determination of appropriate oligonucleotide and lipid concentrations

Other transfection reagents only work in a narrow range based on the final charge ratio of the lipid/DNA complex. GS3815 cytofectin, however, works across a much broader range of charge ratios. We suggest titrating the GS3815 cytofectin and oligonucleotide within the ranges shown below. A good starting point is 2.5 µg/ml of GS3815 cytofectin and 100 nM of the antisense oligonucleotide

1. Oligonucleotide concentration range: 500 nM–1 nM final concentration depending on the potency of the oligonucleotide used.
       phosphorothioate oligonucleotides: 500 nM–100 nM
       propyne modified oligonucleotides: 30 nM–1 nM
2. Lipid concentration range: 1.0 - 20 µg/ml final concentration.

Plating cells

In 100 mm tissue culture dishes, plate cells the day before the transfection at a cell plating density so they will achieve 60-80% cell confluence on the day of transfection.

Oligonucleotide/lipid complex

1. Vortex oligonucleotides well and dispense into eppendorf tubes at the bench (the oligonucleotides do not need to be kept sterile)
2. Transfer the tubes to a tissue culture hood, and add Opti-MEM to each tube to a final volume of 200 µl (i.e., to 4 µl of a 10 µM oligonucleotide stock, add 196 µl Opti-MEM).
       Note: Opti-MEM media can be replaced by serum-free containing MEM.
3. Aliquot the lipid (in a tissue culture hood) into a 12-well *polystyrene* tissue culture plate. For most cell lines, the optimal concentration of GS3815 cytofectin is 2.5 µg/ml final concentration.
       Note: Use of polystyrene plates is essential as the lipid/oligonucleotide complex binds to polypropylene.

4. Add Opti-MEM to each well to a final volume of 200 µl (i.e., to 10 µl of a 1 mg/ml lipid stock add 190 µl Opti-MEM).

5. Add the 200 µl of oligonucleotide/Opti-MEM mixture in the microcentrifuge tube (1.5 ml) from #2 to the corresponding well in the 12-well polystryrene tissue culture plate containing the lipid/Opti-MEM mixture.

$$
\begin{aligned}
& 200 \text{ µl oligonucleotide/Opti-MEM} \\
&+\ \underline{200 \text{ µl lipid/Opti-MEM}} \\
&=\ 400 \text{ µl}
\end{aligned}
$$

6. Incubate 10 to 15 minutes at room temperature

7. Add 3.6 mls complete growth media (10% FBS and antibiotics do not affect transfection efficiency) and mix well for a total transfection volume of 4 mls. In this example, the final transfection mix (4 ml) contains GS3815 lipid at 2.5 µg/ml and the antisense oligonucleotide at 10 nM final concentration.

Table 1. Determination of appropriate transfection volume

Plate size	Transfection vol	Volume 1 mg/ml Lipid
6-well plate	1 ml	2.5 µl
60 mm plate	2 ml	5 µl
100 mm plate	4 ml	10 µl

Transfection

1. Remove media from cells and replace with the 4 mls of the oligonucleotide/lipid transfection mixture. Do not wash cells with PBS or Tris-buffered saline at any time before or after the addition of the transfection mix as this may affect transfection efficiency.

2. Incubate transfection mixture with cells for 4-6 hours, then add an additional 4 mls of complete media. Do not remove the lipid/oligonucleotide transfection mixture at this time unless intolerable toxicity is observed.

3. Incubate the cytofectin/oligonucleotide complex for 24–48 hours before analyzing cells or preparing protein extracts. The time at which extracts are prepared depends on the half-life of your protein of interest. The $t_{1/2}$ of phosphorothioate oligonucleotides is ~35 hours (Flanagan et al., 1996b).

Table 2. Cell lines transfected successfully with GS3815 cytofectin

Cell Line	Species	Type
Cos-7	African Green Monkey	Kidney
CV-1	African Green Monkey	Kidney
BalbC-3T3	Mouse	Embryo fibroblast
Rat-2	Rat	Embryo
Saos-2	Human	Osteogenic sarcoma
HeLa	Human	Cervical carcinoma
NHDF	Human	Dermal fibroblast
MCF-7	Human	Breast carcinoma
A549	Human	Lung carcinoma
T24	Human	Bladder carcinoma
HCT116	Human	Colon carcinoma
H460	Human	Lung carcinoma
WiDR	Human	Colon carcinoma
HT-29	Human	Colon carcinoma
CasKi	Human	Cervical carcinoma
SiHa	Human	Cervical carcinoma

Comments

Although GS3815 cytofectin has been used to successfully deliver oligonucleotides to a wide range of different cell lines, some cell lines are recalitrant to oligonucleotide delivery using any commercially available lipid-mediated product. For the researcher this is a frustrating problem since it is difficult to determine whether the lack of antisense activity is due to targeting a highly structured region of the mRNA sequence or whether they are simply not entering the cell. To address this problem, many researchers include a transfection control in which a fluorescently-labeled oligonucleotide is delivered under identical conditions to a control plate of cells. The cells are then viewed 6-24 h after transfection using a fluorescence microscope and scored for nuclear fluorescence. Cells that are transfected demonstrate bright nuclear fluorescence due to the accumulation of fluorescent oligonucleotide in the nucleus. Cells lines that fail to transfect using cationic lipids may be encouraged to take up oligonucleotides using a different method, such as electroporation.

Table 3. Cell lines transfected with limited success

Cell Line	Species	Type
MDA-MB268	Human	Breast
HL-60	Human	Promyelocytic leukemia
MES	Human	Ovarian carcinoma

6.5 ELECTROPORATION OF OLIGONUCLEOTIDES INTO CELLS

Electroporation is another method by which oligonucleotides can be introduced into cells (Flanagan et al., 1997; Bergan et al., 1993). The high-voltage electric field delivered to the cells causes the temporary disruption of the cellular membrane and allows the entry of oligonucleotides into the cells.

Electroporation procedure

1. Cells are collected (following normal cell trypsinization or centrifugation conditions), rinsed once in complete media to remove any residual trypsin, centrifuged again, and resuspended in complete media to a concentration of 1–3×10^7 cells/ml.
2. The cells (300 µl) are incubated for 10 min at room temperature with oligonucleotides at a final concentration of between 1 and 10 µM. The total volume of the cell suspension and oligonucleotide should not exceed 350 µl.
3. The cell/oligonucleotide mixture is then transferred to a 0.4 cm gap cuvette (Bio-Rad Gene Pulser Cuvette, Cat. #165-2088) and connected to the power supply (Bio-Rad). The power supply delivers a high-voltage pulse of defined magnitude and length. For many non-adherent cells settings of 300 volts (V) and 960 µFarad (µF) have been successfully used although a voltage titration should be done to determine the optimal voltage and capacitance settings.
4. Following delivery of the electrical charge, the cuvette is tapped several times to equilibrate the pH gradient that forms due to electrolysis. The cuvette is then transferred to ice where the cells are allowed to recover for 10 min.
5. The cells are then resuspended in 10 ml of media and plated into a single well of a six-well plate. Alternatively, the sample can be evenly divided between two wells of a six well plate. The cells are then harvested 24–48 h later and processed for RNA or protein extracts.

Comments

Electroporation is an effective method to introduce antisense oligonucleotides into cells although it has several limitations (Flanagan et al., 1997). First, 100-1000 fold more oligonucleotide is required for each sample as compared to a cationic lipid transfection. Second, a large number of cells are required per transfection sample since the electrical pulse required to destabilize the membrane also results in the death of 20-50% of the cells in the cuvette. However, for cells that are not easily transfected with cationic lipids, electroporation may be the only alternative.

6.6 CONCLUSIONS

Technological improvements including the incorporation of nuclease-resistant backbone modifications into oligonucleotides, the discovery of base-modifications that enhance the affinity of oligonucleotides for their target, and the development of cationic lipids to efficiently deliver oligonucleotides to a wide range of cells have overcome many of the barriers limiting the use of antisense inhibitiors in tissue culture assays. Although researchers still need to conduct rigorous controls as outlined in this chapter to avoid non-specific effects of antisense oligonucleotides, antisense technology is becoming a widely used research tool and will play an increasingly important role in the validation and elucidation of therapeutic targets identified by genomic sequencing efforts.

Acknowledgements: I would like to thank present and former members of the code blocker biology group for compiling and editing our laboratory protocols.

6.7 REFERENCES

Azad RF, Driver VB, Tanaka K, Crooke RM, Anderson KP. Antiviral activity of a phosphorothioate oligonucleotide complementary to RNA of the human cytomegalovirus major immediate-early region. *Antimicrob Agents Chemother* 1993;37:1945-54

Benimetskaya L, Loike JD, Khaled Z, Loike G, Silverstein SC, Cao L, El Khoury J, Cai T-Q, Stein CA. Mac-1 (CD11b/CD18) is an oligonucleotide-binding protein. *Nat Med* 1997;3:414-20

Bennett CF, Chiang MY, Chan H, Shoemaker JE, Mirabelli CK. Cationic lipids enhance cellular uptake and activity of phosphorothioate antisense oligonucleotides. *Mol Pharmacol* 1992;41:1023-33

Bergan R, Connell Y, Fahmy B, Neckers L. Electroporation enhances c-myc antisense oligonucleotide efficacy. *Nucleic Acids Res* 1993;21:3567-73

Burgess TL, Fisher EF, Ross SL. The antiproliferative activity of c-myb and c-myc antisense oligonucleotides in smooth muscle cells is caused by a nonantisense mechanism. *Proc Natl Acad Sci USA* 1995;92:4051-55

Coats S, Flanagan WM, Nourse J, Roberts JM. Requirement of p27Kip1 for restriction point control of the fibroblast cell cycle. *Science* 1996;272:877-80

Cowsert LM. In vitro and in vivo activity of antisense inhibitors of ras: potential for clinical development. *Anti-cancer Drug Design* 1997;12:359-71

Dean NM, McKay R, Miraglia L, Geiger T, Muller M, Fabbro D, Bennett CF. Antisense oligonucleotides as inhibitors of signal transduction: development from research tools to therapeutic agents. *Bioch Soc Trans* 1996;24:623-29

Fisher TL, Terhorst T, Cao X, Wagner RW. Intracellular disposition and metabolism of fluorescently-labeled unmodified and modified oligonucleotides microinjected into mammalian cells. *Nucleic Acids Res* 1993;21:3857-65

Flanagan WM, Su LL, Wagner RW. Elucidation of gene function using C-5 propyne antisense oligonucleotides. *Nature Biol* 1996a;14:1139-45

Flanagan WM, Kothavale A, Wagner RW. Effects of oligonucleotide length, mismatches, and mRNA levels on C-5 propyne-modified antisense potency. *Nucleic Acids Res* 1996b;24:2936-41

Flanagan WM, Wagner RW. Potent and selective gene inhibition using antisense oligodeoxynucleotides. *Mol Cel Biochem* 1997;172:213-25

Froehler BC, Wadwani S, Terhorst TJ, Gerrard SR. Oligodeoxynucleotides containing C-5 propyne analogs of 2'-deoxyuridine and 2'-deoxycytidine. *Tetrahedron Lett* 1992;33:5307-10

Graessmann M, Graessmann A. *Microinjection of RNA and DNA into Somatic Cells,* in *Cell Biology: a laboratory handbook,* 1994. JE Celis, Editor. Academic Press, San Diego

Guvakova MA, Yakubov LA, Vlodavsky I, Tonkinson JL, Stein CA. Phosphorothioate oligodeoxynucleotides bind to basic fibroblast growth factor, inhibit its binding to cell surface receptors, and remove it from low affinity binding sites on extracellular matrix. *J Biol Chem* 1995;270:2620-27

Hanvey JC, Peffer NJ, Bisi JE, Thomson SA, Cadilla R, Josey JA, Hassman CF, Bonham MA, Au KG, Carter SG, Bruckenstein DA, Boyd AL, Noble SA, Babiss LE. Antisense and antigene properties of peptide nucleic acids. *Science* 1992;258:1481-85

Ho SP, Bao Y, Lesher T, Malhotra R, Ma LY, Fluharty SJ, Sakai RR. Mapping of RNA accessible sites for antisense experiments with oligonucleotide libraries. *Nature Bio* 1998;16:59-63

Krieg AM, Yi AK, Matson S, Waldschmidt TJ, Bishop GA, Teasdale R, Koretzky GA, Klinman DM. CpG motifs in bacterial DNA trigger direct B-cell activation. *Nature* 1995;374:546-49

Lewis JG, Lin KY, Kothavale A, Flanagan WM, Matteucci MD, DePrince RB, Mook RA, Hendren RW, Wagner RW. A serum-resistant cytofectin for cellular delivery of antisense oligodeoxynucleotides and plasmid DNA. *Proc Natl Acad Sci USA* 1996;93:3176-81

Matteucci MD, Wagner RW. In pursuit of antisense. *Nature* 1996;384:suppl. 20-22

Monia BP, Johnston JF, Greiger T, Muller M, Fabbro D. Antitumor activity of a phophorothioate antisense oligodeoxynucleotide targeted against c-raf kinase. *Nat Med* 1996a;2:668-75

Monia BP, Sasmor H, Johnston JF, Freier SM, Lesnk EA, Muller M, Altmann K-H, Moser H, Fabbro D. Sequence specific antitumor activity of a phosphorothioate oligodeoxyribonucleotide targeted to human c-raf kinase supports an antisense mechanism of action in vivo. *Proc Natl Acad Sci USA* 1996b;93: 15481-85

Monia BP. First- and second-generation antisense inhibitors targeted to human c-raf kinase: in vitro and in vivo studies. *Anti-cancer Drug Des* 1997;12:327-41

Moulds C, Lewis JG, Froehler BC, Grant D, Huang T, Milligan JF, Matteucci MD, Wagner RW. Site and mechanism of antisense inhibition by C-5 propyne oligonucleotides. *Biochemistry* 1995;34:5044-53

Patzel V, Sczakeil G. Theorectical design of antisense RNA structures substantially improves annealing kinetics and efficacy in human cells. *Nature Bio.* 1998;16:64-68

Shaw JP, Kent K, Bird J, Fishback J, Froehler B. Modified deoxyoligonucleotides stable to exonuclease degradation in serum. *Nucleic Acids Res* 1991;19:747-50

St. Croix B, Floerenes VA, Rak J, Flanagan WM, Kerbel RS. Impact of p27kip1 on adhesion-dependent resistance of tumor cells to anti-cancer agents. *Nat Med* 1996;2:1204-10

Stein CA, Krieg AM. Problems in interpretation of data derived from *in vitro* and *in vivo* use of antisense oligodeoxynucleotides. *Antisense Res Dev* 1994;4:67-69

Stein CA. Does antisense exist? *Nat Med* 1995;1:1119-21

Stein CA. Phosphorothioate antisense oligonucleotides: questions of specificity. *Trends Biotechnol* 1996;14:147-49

Wagner RW, Matteucci MD, Lewis JG, Gutierrez AJ, Moulds C, Froehler BC. Antisense gene inhibition by oligonucleotides containing C-5 propyne pyrimidines. *Science* 1993;260:1510-13

Wagner RW. Gene inhibition using antisense oligodeoxynucleotides. *Nature* 1994;372:333-35

Wagner RW. The state of the art in antisense research. *Nat Med* 1995a;1:1116-18

Wagner RW. Toward a broad-based antisense technology. *Antisense Res Dev* 1995b;5:113-14

Wagner RW, Flanagan WM. Antisense technology and prospects for therapy of infectious disease and cancer. *Mol Med Tod* 1997;3:31-38

Zamecnik PC, Stephenson ML. Inhibition of Rous sarcoma virus replication and cell transformation by a specific oligodeoxynucleotide. *Proc Natl Acad Sci USA* 1978;75:280-84

7 Inhibition of TNF Synthesis by Antisense Oligonucleotides

Margaret Taylor* and Lester Kobzik,

Harvard School of Public Health, Boston, MA, USA,
*current affiliation Sequitur, Inc., Waltham, MA, USA

7.1 INTRODUCTION

Tumor necrosis factors (TNF) are a family of proinflammatory cytokines consisting of two members, TNF-α and lymphotoxin, which mediate a number of inflammatory and immune responses (Beutler et al., 1989). TNF-α is produced by a number of cell types including monocytes/ macrophages, mast cells, lymphocytes, and keratinocytes. TNF-β is produced primarily by activated lymphocytes. TNF is released early in the inflammatory cascade and triggers a wide range of cellular events (Leeper-Woodford et al., 1991) including the induction of interleukin-1, interleukin-8, and platelet activating factor. TNF also causes upregulation of adhesion molecules on neutrophils and endothelial cells, enhancement of vascular permeability (Beutler et al., 1989; Leeper-Woodford et al., 1991; Tracey et al., 1988), and stimulation of neutrophil respiratory burst activity (Nathan, 1987).

Uncontrolled production of TNF has been implicated in the pathogenesis of a number of diseases including, rheumatoid arthritis (Maini et al., 1997; Moreland et al., 1997), multiple sclerosis (Shareif et al., 1991), and adult respiratory distress syndrome (Hyers et al., 1991; Roten et al., 1991; Roumen et al., 1993; Headley et al., 1997). Inhibition of TNF by neutralizing antibodies and/or soluble TNF receptor has been shown to abrogate its deleterious effects in several models of inflammation (Warren et al., 1989; Tracey et al., 1990; Vogel et al., 1990; Denis et al., 1991; Ulich et al., 1993). The use of antisense oligomers targeted toward TNF represents an alternative intervention to attenuate uncontrolled inflammatory responses.

Phosphorothioate oligonucleotides inhibit the production of human TNF by activated U937 cells (Yahata et al., 1996; Lefebvre d'Hellencourt et al., 1997) and peripheral blood mononuclear cells (Hartmann et al., 1996). In addition, phosphorothioate-mediated inhibition of TNF production by activated lymphocytes has been shown (Lefebvre d'Hellencourt et al., 1997; Taylor et al., 1997). Rojanasakul et al. demonstrated inhibition of silica-induced TNF production by rat alveolar macrophages only when phosphorothioate oligonucleotides were conjugated to the drug targeting vector mannosylated polylysine (Rojanasakul et al., 1997). We have demonstrated partial inhibition of lipopolysaccharide induced TNF by RAW 264.7 cells and by murine alveolar macrophages.

The following chapter will detail some of the experiments in which antisense oligomers have been used to inhibit TNF production. The effects of antisense agents in three cell types will be discussed.

7.2 METHODS AND RESULTS

Description of antisense oligomers

Morpholino oligomers used in these studies were provided by AntiVirals, Inc. (Corvallis, OR), and were synthesized by methods described previously (Summerton, 1993; Summerton et al., 1993). Morpholino oligomers are a new generation of antisense agents in which the ribose moiety has been altered to a six membered ring, containing a nitrogen atom. These oligomers have been shown highly specific and nuclease resistant (Hudziak et al, 1996). Antisense effects of morpholino oligomers are thought to occur via translational arrest as morpholino-RNA duplexes are not substrates for RNAse H. In the studies described below, we compared the efficacy of morpholino oligomers with that of phosphorothiote oligonucleotides which exert antisense effects via activation of RNAse H. We screened a panel of oligomers (Table 1) in order to establish a rank order of efficacy.

Table 1: Description of antisense sequences and target sites

Chemistry	Name	length	Position of target site	Sequence (5'→ 3')
Morpholino	M-AS 1	20	AUG start codon (-13- +7)	GCU CAU GGU GUC UUU UCU GG
	M-AS 2	24	AUG start codon (-5- +19)	CAU GCU UUC UGU GCU CAU GGU GUC
	M-AS 3	19	AUG start codon (-12- +7)	GCU CAU GGU GUC UUU UCU G
	M-AS 4	22	5' UTR (-85- -63)	GGU UGG CUG CUU GCU UUU CUG G
	M-NS 1	20	None	AAG CGC CAA UGA GUU GAC UC
	M-NS 2	24	None	UCG GCU UCG CGC AGU UAU CUC UUU
Methyl-morpholino*	MAS2me	24	AUG start codon (-5- +19)	MAT GMT TTM TGT GMT MAT GGT GTM
	MAS5me	21	AUG start codon (-5- +16)	GMT TTM TGT GMT MAT GGT GTM
Phospho-rothioate	S-AS 2	24	AUG start codon (-5- +19)	CAT GCT TTC TGT GCT CAT GGT GTC
	S-NS 1	20	None	AAG CGC CAA TGA GTT GAC TC
	S-NS 2	24	None	TCG GCT TCG CGC AGT TAT CTC TTT

* Oligomers have 5' methyl-modification on the base, M= methyl C, T= methyl U

Inhibition of TNF production by RAW 264.7 cells

We have shown that morpholino oligomers (M-AS), but not phosphorothioate oligonucleotides, partially inhibit lipopolysaccharide (LPS)-induced TNF production by macrophages (Taylor et al., 1996; Taylor et al., 1997; Taylor et al., 1998).

Treatment of cells with antisense oligonucleotides

RAW 264.7 cells were obtained from the American Type Culture Collection and were cultivated in RPMI-1640 supplemented with 10% fetal bovine serum (FBS), 2mM L-glutamine, 100 units/ml penicillin, and 0.1mg/ml streptomycin. Cells were maintained at 37°C with 5% CO_2 in a humidified incubator.

In order to maximize the reproducibility of antisense effects, the cells were maintained on a schedule before use. Specifically, two days prior to use in the assay, cells were seeded at 6 x 10^6 in 20 ml in a 100 mm tissue culture plate. The next day, cells were plated at 2 x 10^5 cells / well (100 ml volume) in a 96 well format. On the day of treatment, antisense oligomers were mixed with lipofectin (10 μg/ml, see below) at room temperature for 15 minutes. The oligo-lipofectin mixtures were added to the cells and were incubated for 4 or 6 hours at 37°C. After this pre-incubation, cells were incubated with 37.5 ng/ml LPS in the presence of 0.5% FBS for an additional 4 hours. This dose of LPS was chosen after a dose response analysis because it caused approximately 60 % of maximal TNF release. Upon completion of the incubation, cell supernatants were harvested and their TNF-alpha content was determined by a specific ELISA.

Lipofectin is a cationic lipid preparation which has been shown to improve the uptake of negatively charged molecules by cells. We included with our M-AS (charge neutral) as a control for incubation with S-AS (negatively charged). Surprisingly, we found that lipofectin improved the reproducibility of morpholino antisense effects. The mechanism by which this improvement occurs remains unclear. The mechanism may be due to a direct effect of lipofectin (e. g. toxicity) on cell function and may not be mediated by improved oligonucleotide delivery.

Of the four specific oligomers tested, only M-AS 2 caused significant inhibition of TNF production by RAW 264.7 cells. The inhibition was specific and dose dependent (Figure1). In this system, maximal inhibition of 32.6 +/- 2.6 % was obtained when cells were incubated with 25 μM M-AS 2 for 6 hours prior to stimulation by LPS. Under these conditions, none of the other oligomers tested caused significant inhibition (Figure 1A).

A.

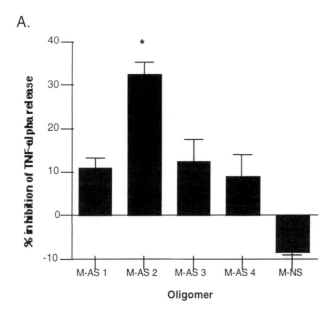

B.

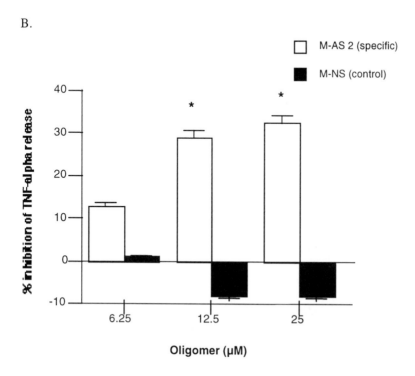

Figure 1:Effect of morpholino oligomers on LPS-induced TNF-α production by RAW 264.7 cells is sequence specific and dose-dependent. (A)RAW 264.7 cells were pre-treated with 25 μM morpholino antisense oligomers in the presence of 10μg/ml lipofectin for 6 hours, then stimulated with 37.5 ng/ml LPS, and after 4 hours TNF-α in cell supernatants was quantitated by ELISA. M-AS 2 caused greater inhibition of TNF-α production than did the other morpholino oligomers tested. Results shown represent the mean % inhibition of TNF-α production ± SEM, n≥3 experiments performed in triplicate. TNF-α produced in samples treated with lipofectin only was used to establish the 0% inhibition line. * indicates statistically significant inhibition p< 0.02. (B) RAW 264.7 cells were pre-incubated with either M-AS 2 (white bars) or M-NS (black bars) (25 μM, 12.5 μM, or 6.25 μM) for 6 hours in the presence of 10 μg/ml lipofectin, then stimulated with 37.5 ng/ml LPS, and after 4 hours TNF-α in cell supernatants was quantitated by ELISA. Results shown represent the mean % inhibition of TNF-α production ± SEM, n≥5 experiments performed in triplicate. TNF-α produced in samples treated with lipofectin only was used to establish the 0% inhibition line. * indicates statistically significant inhibition p< 0.001.

We found that S-AS did not inhibit TNF production by RAW 264.7 cells. In fact, we observed augmentation of the TNF response in the presence of the antisense sequence, S-AS 2, which was targeted to the same region of the TNF sequence as was M-AS 2 (Table 2). We also observed this augmentation upon incubation with the mismatch control sequence S-NS. Hartmann et al. also reported that phosphorothioate oligonucleotides augment LPS-induced TNF production in mononuclear cells (Hartmann et al., 1996). They further demonstrated that the augmentation by oligonucleotides was reversible by heparin-treatment, suggesting that the amplification of TNF synthesis was caused by interaction of the polyanionic oligonucleotides with cationic sites on the cell surface (Hartmann et al., 1996). Immunostimulatory properties of S-AS containing CpG motifs have also been reported (Krieg et al., 1995; Krieg, 1996). The control sequence contained one CpG motif which might explain its augmentation of TNF production by RAW 264.7 cells; however, the antisense sequence, S-AS 2, contained no CpG motifs. Relatively high concentrations of the oligonucleotide (> 12.5 μM) were necessary to achieve significant inhibition of TNF synthesis by morpholino antisense oligonucleotides. However, all other sequences tested under the same conditions did not reduce TNF synthesis suggesting that TNF inhibition was not due to the high oligonucleotide concentration

In conclusion we showed that morpholino, but not phosphorothioate antisense sequences are able to inhibit LPS-induced TNF by RAW 264.7 macrophages.

Inhibition of TNF production by T-lymphocytes

HT2 cells, a murine T-lymphocyte cell line, were obtained from the American Type Culture Collection and were cultivated in RPMI-1640 supplemented with 10% fetal bovine serum (FBS), 2mM L-glutamine, 100 units/ml penicillin, 0.1mg/ml streptomycin, 50 U/ml murine IL-2, and 0.05 mM 2-mercaptoethanol. Cells were maintained at $37^{\circ}C$ with 5% CO_2 in a humidified incubator.

Treatment of cells with antisense

One day prior to use in the assay, 4×10^6 HT2 cells were seeded in RPMI-1640 media with 10 U/ml IL-2 in a 100 mm tissue culture dish. On the day of treatment, cells were plated at 5×10^4 per well in 96-well plates, and were incubated with antisense oligomers (phosphorothioate or morpholino) at the concentrations indicated below. Antisense agents were added to the cells in the presence or absence of lipofectin (10 μg/ml).

Antisense sequences were mixed with lipofectin or RPMI-1640 for 15 minutes at room temperature, added to the HT2 cells, and the cells were incubated at $37^{\circ}C$ for 4 hours in serum-free RPMI. After preincubation, cells were stimulated with 12.5 nM phorbol myristate acetate (PMA) and were incubated for an additional 4 h in a final concentration of 5.0 % serum. Upon completion of the incubation, cell supernatants were harvested and their TNF-α protein levels were quantitated by specific ELISA (Endogen).

Effects of phosphorothioate oligonucleotides

HT2 cells were treated with 0.5 µM, 0.25 µM, or 0.13 µM S-AS 2 and S-NS in the presence or absence of lipofectin (10 µg/ml). In the absence of lipofectin, S-AS 2 significantly inhibited TNF production at all concentrations tested (31.4 % ± 1.2%, 27.9 % ± 4.2 %, 22.6 % ± 4.0%; p= 0.0002, 0.04, and 0.05, respectively). The nonsense control oligonucleotide S-NS did not cause significant inhibition of TNF at these concentrations (Figure 2).

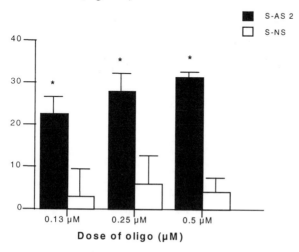

Figure 2: Dose-response analysis of inhibition of TNF-α release by the phosphorothioate oligonucleotide, S-AS 2. HT2 cells were pre-incubated with either S-AS 2 (black bars) or S-NS (white bars) for four hours at the concentrations indicated. The cells were then stimulated with 12.5 nM PMA (HT2 cells) or 37.5 ng/ml LPS (RAW 264.7 cells) and incubated an additional four hours. Subsequently, TNF-α levels in cell supernatants were quantitated by ELISA. Results shown represent the mean % inhibition of TNF-α release ± SEM n=3 experiments performed in duplicate. TNF-α release measured in samples treated with media alone was used to define 0 % inhibition. * indicates statistically significant inhibition compared with the media control, p=0.05. y-axis: Percent inhibition of TNF-α release.

Lipofectin is a cationic lipid preparation which has been shown to enhance the delivery and efficacy of phosphorothioate oligonucleotides (Barbour et al., 1993; Bennett et al. 1994). Surprisingly, we observed that the efficacy of S-AS 2 in HT2 cells was significantly diminished in the presence of 10 µg/ml lipofectin (Figure 3). Using FITC-labeled oligonucleotides, we showed that the uptake of 0.5 µM FITC-S-AS 2 and FITC-S-NS was not improved by the addition of lipofectin (Table 2). We hypothesize that lipofectin may alter the mechanism by which oligonucleotides are taken up by cells, thus altering their efficacy. Specifically, lipofectin may cause the formation of oligonucleotide lipid aggregates resulting in their uptake by a phagocytic rather than endocytic (e.g. adsorptive) mechanism.

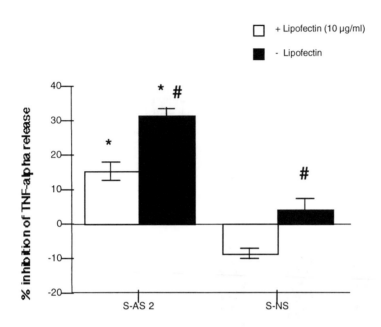

Oligonucleotide (0.5 µM)

Figure 3: The efficacy of phosphorothioate oligonucleotides is diminished in the presence of lipofectin. HT2 cells were pre-incubated with either S-AS 2 or S-NS (0.5 µM) in the presence (white bars) or absence (black bars) of lipofectin (10 µg/ml) for four hours, then stimulated with 12.5 nM PMA, and after four hours, TNF-α in cell   supernatants was quantitated by ELISA. Results shown represent the mean % inhibition ± SEM, n=3 experiments performed in duplicate. Control samples which were not treated with oligo were used to define 0% inhibition.* indicates significant inhibition compared with control, # indicates that lipofectin treatment is significantly different from naked oligonucleotide treatment, p<0.05.

Table 2: Measurement of uptake of FITC-labeled phosphorothioate oligonucleotides

Oligonucleotide	Lipofectin	Mean fluorescence (flow cytometry)
S-AS 2	+	94.4 +/- 4
S-NS 1	+	94 +/- 5.9
S-AS 2	-	111.2 +/- 4.5
S-NS 1	-	88.6 +/- 2.8

HT2 cells were incubated with 0.5 µM FITC-labeled phosphorothioate oligonucleotides in the presence or absence of lipofectin for 4h. Following incubation, the cells were harvested and cell associated fluorescence was measured by flow cytometry.

Effects of morpholino oligomers

As described above, we showed that morpholino-type oligomers significantly inhibited TNF production by LPS-stimulated RAW 264.7 cells. We predicted that the efficacy of these oligomers would be improved in a cell line with a less well-developed endosomal-lysosomal network. Toward testing this hypothesis, we evaluated the efficacy of morpholino-modified oligomers in HT2 cells. Contrary to our prediction, we found that the efficacy of 25 µM M-AS 2 was diminished in HT2 cells (Figure 4) compared with RAW 264.7 cells (12.3% vs. 32%).

Morpholino oligomers are uncharged molecules; thus there is no physico-chemical rationale for an interaction between these oligomers and lipofectin, a cationic lipid preparation. Originally we included lipofectin in the morpholino incubations in order to allow a fair comparison between the phosphorothioate and morpholino oligomers. Surprisingly, we found that in the absence of lipofectin, the efficacy of morpholino oligomers was significantly decreased (Figure 4). Further, we showed, using FITC-labeled oligomers, that the uptake of 25 µM M-AS 2 and M-NS was improved by incubation with lipofectin (Table 3).

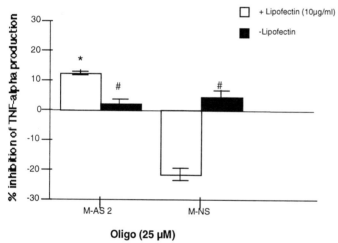

Oligo (25 µM)

Figure 4: The efficacy of morpholino oligomers is improved by the addition of lipofectin. HT2 cells were pre-incubated with either M-AS 2 or M-NS in the presence (white bars) or absence (black bars) of lipofectin (10 µg/ml) for four hours, then stimulated with 12.5 nM PMA, and after 4 hours, TNF-α levels in cell supernatants were quantitated by ELISA. Results shown represent the mean % inhibition ± SEM, n=2 experiments performed in duplicate. Control samples which were not treated with oligomer were used to define 0% inhibition.* indicates significant inhibition compared with control (no oligomer), # indicates significant difference between lipofectin treatment and naked oligomer treatment, p<0.02.

In summary, these studies indicated that the efficacy of antisense sequences is influenced by both the chemical composition of the antisense as well as the choice of target cell type. Our work indicates that antisense agents targeted toward cytokines may be useful. Phosphorothioate oligonucleotides seem better suited toward lymphocyte targets, while morpholino oligomers show promise toward macrophage targets. Improvement of delivery to the target site remains an obstacle for effective antisense inhibition in both cell types.

Effect of antisense in primary AMs

To determine whether antisense oligomers could be effective in primary cells, we investigated the efficacy of S-AS and M-AS in primary murine alveolar macrophages (AMs). We predicted that the effects of antisense in primary macrophages would be similar to those observed in the RAW 264.7 cells.

Preparation of cells and treatment with antisense

AMs were obtained from Balb/c mice by bronchoalveolar lavage with phosphate buffered saline (PBS)-EDTA. After centrifugation and resuspension, the cells were counted, adjusted to $1X\ 10^6$ cells/ml, and used immediately. Viability of recovered cells was determined by trypan blue exclusion and was greater than 95% in all lavages.

AMs were plated in 96-well low binding cell culture plates (Costar, Cambridge, MA) at $5\ X\ 10^4$/well. Antisense (AS) or nonsense (NS) oligomers at the indicated concentrations were mixed with lipofectin (10 µg/ml) (LifeTechnologies, Inc.) in serum-free RPMI for 15 minutes at room temperature. The oligomer, oligomer-lipofectin mixture, or lipofectin only was added to the AMs and the cells were incubated for 2 hours in serum-free conditions at $37^{\circ}C$. After this pre-incubation, AMs were stimulated with 50 ng/ml LPS, and incubated for an additional 3 hours. Following incubation, cell supernatants were collected and TNF levels were determined by specific ELISA (Endogen). Cell viability was determined after treatment by trypan blue exclusion. Average cellular viability was 83.6 +/- 6.6% and did not differ among the treatment groups.

In this system, adherence to a solid substrate was also used as a stimulus for TNF production. Adherence to plastic mimics the adherence of macrophages to the extracellular matrix during tissue injury. AMs obtained as described above were plated in 96 well plates at $5\ X\ 10^4$. Cells were treated with antisense or controls as described and were incubated for 3 hours at $37^{\circ}C$. During the 3 hour incubation, the cells adhered to the tissue culture plastic. Following incubation, cell supernatants were harvested and TNF was quantitated by ELISA.

In these experiments, two (2) additional morpholino oligomers were tested for efficacy. These oligomers contained methylated cytosine and uracil residues. Methyl-modification of these bases is reported to increase the affinity of the oligomer for the substrate (Grigoriev, et al., 1992). We evaluated two methyl-modified sequences, M-AS 2me and M-AS 5 me, and found that their efficacy was not significantly improved compared to that of M-AS 2 (See results below).

As we predicted, the efficacy of morpholino antisense oligomers was similar in primary AMs as it was in RAW 264.7 cells. M-AS 2, M-AS 2me and M-AS 5me (25 μM) significantly inhibited TNF production by AMs stimulated by LPS (36.6 +/- 3.2, 27.3 +/- 3.0, and 37.7 +/- 1.9% inhibition, respectively), while the nonsense control oligomers were without effect (Figure 5A). Increasing the concentration of oligomer to 50 μM significantly improved the efficacy of M-AS 2 me to 45.9 +/- 3.7 % vs. 27.3 +/- 3.0% inhibition, $n \geq 3$, p=0.005), but only slightly improved the efficacy of M-AS 2 (Figure 5B).

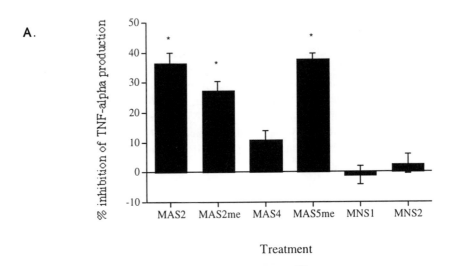

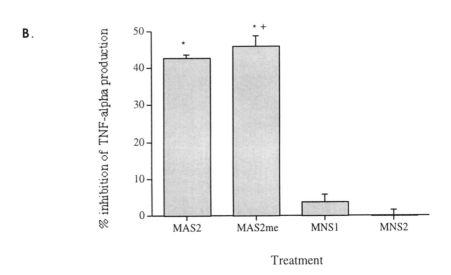

Figure 5: Morpholino antisense oligomers inhibit LPS-induced TNF-α production by murine alveolar macrophages. (A) AMs were pretreated with 25μM morpholino antisense oligomers in the presence of 10 μg/ml lipofectin for 2 h, then stimulated with 50 ng/ml LPS. After 4 h, TNF-α in cell supernatants was quantitated by ELISA. M-AS 2, M-AS 2me, and M-AS 5 significantly inhibited TNF-α production. M-AS 4 caused slight, but not statistically significant inhibition. The nonsense oligomers M-NS 1 and M-NS 2 did not inhibit TNF-α production. Results shown represent the mean percent inhibition ± SEM, n=3 experiments performed in duplicate. TNF-α produced by cells treated with lipofectin only was used to establish 0% inhibition. * indicates statistcally significant inhibition, p< 0.0001. (B) Using the above protocol, a subset of the oligomers were evaluated at a concentration of 50 μM. M-AS 2 and M-AS 2me caused significant inhibition of TNF-α production. Results shown

represent the mean percent inhibition ± SEM, n=3 experiments performed in duplicate. * indicates statistcal differences from lipofectin treatment group, p< 0.0001. + indicates significantly greater inhibition by 50 µM than by 25 µM M-AS 2me.

Adherence of macrophages to a solid substrate results in the production of TNF, a phenomenon which may contribute to macrophage activation when tissue matrix is exposed by injury. We used adherence as a stimulus to evaluate the ability of four morpholino oligomers targeted against TNF and two nonsense control oligomers to inhibit TNF production. Incubation of AMs in the presence of M-AS 2 and M-AS 2me (25 µM) significantly reduced TNF production (28.7 +/- 2.2 % and 29.4 +/- 8.2% inhibition, respectively) (Figure 6A).

Dose response analysis of a subset of the oligomers (Figure 6B) demonstrated that inhibition of TNF by M-AS 2 was significantly improved at 50 µM compared with 25 µM (52.6 +/- 0.74 % vs. 29.6 +/- 2.8 %, n=3, p=0.002). Control (nonsense) oligomers were without effect at either 25 µM or 50 µM. In one experiment using 100 µM concentration of the oligomers, we demonstrated complete inhibition of TNF by both M-AS 2 and M-AS 2me; however, at that concentration, nonsense oligomers inhibited TNF production by approximately 50%.

A.

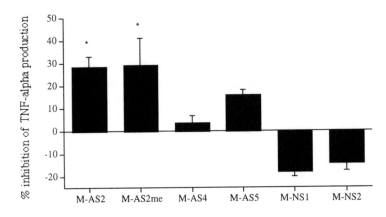

Treatment

B.

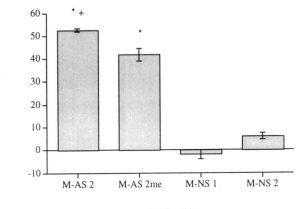

treatment

Figure 6: Morpholino antisense oligomers inhibit adherence-induced TNF-α production by murine alveolar macrophages. (A) AMs were pretreated with 25 μM morpholino antisense oligomers in the presence of 10 μg/ml lipofectin for 2 h, then stimulated with 50 ng/ml LPS. After 4 h, TNF-α in cell supernatants was quantitated by ELISA. M-AS 2 (n=5) and M-AS 2me (n=2) significantly inhibited TNF-α production. M-AS 4 (n=2) and the nonsense oligomers M-NS 1 (n=3) and M-NS 2 (n=4) did not inhibit TNF-α production. Results shown represent the mean percent inhibition ± SEM, experiments performed in duplicate. TNF-α produced by cells treated with lipofectin only was used to establish 0% inhibition. * indicates statistcally different from lipofectin treatment group, p< 0.02. **(B)** Using the above protocol, a subset of the oligomers were evaluated at a concentration of 50 μM. M-AS 2 and M-AS 2me caused significant inhibition of TNF-α production. Results shown represent the mean percent inhibition ± SEM, n=3 experiments performed in duplicate. * indicates statistcal difference from lipofectin treatment group, p< 0.0001. + indicates significantly greater inhibition by 50 μM than by 25 μM M-AS 2.

7.3 CONCLUDING REMARKS

The results presented here and detailed in (Taylor et al., 1996; Taylor et al., 1997; Taylor et al., 1998) demonstrate that antisense oligomers can inhibit TNF production by macrophages and T-lymphocytes. Our results indicate that phosphorothioate oligonucleotides are more efficacious in lymphocytes, while morpholino oligomers show greater potential in macrophages. We achieved partial inhibition of TNF in RAW, HT2 and primary AMs. Inhibition of TNF production using phosphorothioate antisense oligonucleotides has also been demonstrated in activated monocytes and lymphocytes (Hartmann, et al, 1996a, Yahata, et al., 1996, Lefebvre d'Hellencourt, et al., 1997). Adequate delivery to the target site remains an obstacle to the efficacy of these antisense agents. We conclude that antisense oligomers targeted toward cytokines have potential utility for discerning gene function in vitro; however, further experiments will be required to assess whether antisense oligomers can effectively regulate cytokine production in vivo.

7.4 REFERENCES

Barbour SE, Dennis EA. Antisense inhibition of group II phospholipase A2 expression blocks the production of prostaglandin E2 by P388D1 cells. *J Biol Chem* 1993;268:21875-82

Bennett CF, Condon TP, Grimm S, Chan H, Chiang M. Inhibition of endothelial cell adhesion molecule expression with antisense oligonucleotides. *J Immunol* 1994;152:3530-40

Beutler B, Cerami A. The biology of cachectin/ TNF-α primary mediator of the host response. *Ann Rev Immunol* 1989;7:625-55

Denis M, Cormier Y, Fournier M, Tardif J, Laviolette M. Tumor necrosis factor plays an essential role in determining hypersensitivity pneumonitis in a mouse model. *Am J Resp Cell Mol Biol* 1991;5:477-83

Hartmann G, Krug A, Eigler A, Moeller J, Murphy J, Albrecht R, Endres S. Specific suppression of human tumor necrosis factor-α synthesis by antisense oligodeoxynucleotides. *Antisense Nucl Acid Drug Dev* 1996a;6:291-99

Hartmann G, Krug A, Waller-Fontaine K, Endres S. Oligodeoxynucleotides enhance lipopolysaccharide-stimulated synthesis of tumor necrosis factor: dependence on phosphorothioate modification and reversal by heparin. *Mol Med* 1996b;2:429-38

Headley AS, Tolley E, Meduri GU. Infections and the inflammatory response in acute respiratory distress syndrome. *Chest* 1997;111:1306-21

Hyers TM, Tricomi SM, Dettenmeier PA, Fowler AA. Tumor necrosis factor levels in serum and bronchoalveolar lavage fluid of patients with the adult respiratory distress syndrome. *Am Rev Respir Dis* 1991;144:268-71

Krieg A, Yi A, Matson S, Waldschmidt TJ, Bishop GA, Teasdale R, Koretzky GA, Klinman DM. CpG motifs in bacterial DNA trigger direct B-cell activation. *Nature* 1995;374:546-49

Krieg AM. An innate immune defense mechanism based on the recognition of Cpg motifs in microbial DNA. *J Lab Clin Med* 1996;128:128-33

Leeper-Woodford SK, Carey PD, Byrne K, Jenkins JK, Fisher BJ, Blocher C, Sugerman HJ, Fowler III AA. Tumor necrosis factor: alpha and beta subtypes appear in circulation during onset of sepsis induced lung injury. *Am Rev Respir Dis* 1991;143:1076-82

Lefebvre d'Hellencourt C, Diaw L, Cornillet P, Guenounou M. Inhibition of human TNFα and LT in cell-free extracts and in cell culture by antisense oligonucleotides. *Biochim Biophys Acta* 1996;1317:168-74

Maini RN, Elliott M, Brennan FM, Williams RO, Feldmann M. TNF blockadein rheumatoid arthritis: implications for therapy and pathogenesis. *APMIS* 1997;105:257-63

Moreland LW, Baumgartner SW, Schiff MH, Tindall EA, Fleischmann RM, Weaver AL, Ettlinger RE, Cohen S, Koopman WJ, Mohler K, Widmer MB, Blosch CM. Treatment of rheumatoid arthritis with a recombinant human tumor necrosis factor receptor (p75)-Fc fusion protein. *N Engl J Med* 1997;337:141-47

Nathan CF. Neutrophil activation on biological surfaces: Massive secretion of hydrogen peroxide in response to products of macrophages and lymphocytes. *J Clin Invest* 1987;80:1550-60

Rojanasakul Y, Weissman DN, Shi X, Castranova V, Ma JKH, Liang W. Antisense inhibition of silica-induced tumor necrosis factor in alveolar macrophages. *J Biol Chem* 1997;272:3910-14

Roten R, Markert M, Feihl F, Schaller M, Tagan M, Perret C. Plasma levels of tumor necrosis factor in the adult respiratory distress syndrome. *Am Rev Respir Dis* 1991;143:590-92

Roumen RM, Hendriks T, Ven-Jongekrijg JVD, Nieuwenhuijzen GA, Sauerwein RW, Meer JWVD, Goris RJ. Cytokine patterns in patients after major vascular surgery, hemorrhagic shock, and severe blunt trauma. *Ann Surg* 1993;218:769-76

Sharief MK, Hentges R. Association between tumor necrosis factor and disease progression in patients with multiple sclerosis. *N Engl J Med* 1991;325:467-72

Summerton J. AntiVirals Technical Report #3. *Antisense Res Dev* 1993;3:306-09

Summerton J, Weller D. Uncharged morpholino-based polymers having phosphorous containing intersubunit linkages. 1993; US patent 5,185,444

Taylor M, Paulauskis J, Kobzik L. In vitro efficacy of morpholino-modified antisense oligomers directed against tumor necrosis factor-alpha mRNA. *J Biol Chem* 1996;271:17445-52

Taylor P, Weller D, Kobzik L. Comparison of antisense oligomers directed toward TNF-alpha in helper T and macrophage cell lines. *Cytokine* 1997;9:672-81

Taylor M, Weller D, Kobzik L. Effect of TNF-alpha antisense oligomers on cytokine production by primary murine alveolar macrophages. *Antisense Nucl Acid Drug Dev* 1998;8:199-205

Tracey K, Lowry S, Cerami A. Cachectin/TNF-alpha in septic shock and septic adult respiratory distress syndrome. *Am Rev Respir Dis* 1988;138:1377-79

Tracey KJ, Cerami A. Metabolic responses to cachectin/TNF: a brief review. *Ann NY Acad Sci* 1990;587:325-31

Ulich TR, Yin S, Remick DG, Russell D, Eisenberg SP, Kohno T. Intratracheal administration of endotoxin and cytokines: IV. The soluble tumor necrosis factor type I inhibits acute inflammation. *Am J Pathol* 1993;42:1335-38

Vogel SN, Havell EA. Differential inhibition of lipopolysaccharide-induced pneumonia by anti-tumor necrosis factor alpha antibody. *Infect Immun* 1990;58:1873-82

Warren JS, Yabroff KR, Remick DG, Kunkel SL, Chensue SW, Kunkel RG, Johnson KJ, Ward PA. Tumor necrosis factor participates in the pathogenesis of acute immune complex alveolitis in the rat. *J Clin Invest* 1989;84:1873-82

Yahata N, Kawai S, Higaki M, Mizushima Y. Antisense phosphoothioate oligonucleotide inhibits interleukin 1-beta production in the human macrophage-like cell line, U937. *Antisense Nucl Acid Drug Dev* 1996;6:55-61

8 Inhibition of IGF-I Receptor Expression by Antisense Oligonucleotides in vitro and in vivo

Mariana Resnicoff

Kimmel Cancer Center
Philadelphia, PA, USA

8.1 INTRODUCTION

The insulin-like growth factor I receptor (IGF-IR) belongs to the family of tyrosine kinase receptors and it shares 70% homology with the insulin receptor (Ullrich et al., 1986). The IGF-IR activated by its ligands (IGF-I, IGF-II and insulin, at supraphysiological concentrations) plays a major role in cell proliferation by at least three different mechanisms:
1. it is mitogenic;
2. it is required for the establishment and maintenance of the transformed phenotype in several cell types;
3. it protects cells from apoptosis induced by a variety of stimuli, both *in vitro* and *in vivo*.

The IGF-IR in mitogenesis

The essential role of the IGF system *in vivo* has been demonstrated by Efstratiadis and co-workers who showed that targeted disruption of both the IGF-II and IGF-IR genes (by homologous recombination) results in progeny with a body weight at birth that is 30% that of wild-type littermates (Liu et al., 1993; Baker et al., 1993). Given that IGF-II is the major ligand of the IGF-IR in mouse embryos, it can be concluded that the activated IGF-IR accounts for 70% of embryonic murine growth.

3T3-like fibroblasts (R-) generated from mouse embryos with a targeted disruption of the IGF-IR genes were analyzed for their growth requirements in monolayer cultures. R- cells do not grow in serum-free medium supplemented with purified growth factors that sustain the growth of mouse embryo fibroblasts derived from wild-type littermates (Sell et al., 1993; Sell et al., 1994). However, R- cells are able to grow in medium supplemented with 10% serum, but with a lower rate than wild-type cells.

These observations indicate that the IGF-IR is not essential for growth, although it may be required for optimal cell growth. The IGF-IR is required for optimal growth of several cell types in culture, such as human diploid fibroblasts, endothelial cells, epithelial cells, condrocytes, osteoblasts, T lymphocytes, myeloid cells and stem cells from the bone marrow.

The IGF-IR in transformation

An overexpressed or constitutively active IGF-IR leads to the establishment of a transformed phenotype, as determined by the ability to form colonies in soft agar or develop tumors in nude mice (for a review: Baserga, 1995).

The crucial role of the IGF-IR in transformation was demonstrated in R- cells, which are refractory to transformation by a variety of viral and cellular oncogenes (SV40 T antigen, an activated ras alone or in combination with SV40 T antigen, bovine or human papilloma virus and overexpressed growth factor receptors), all of which are able to transform wild-type cells (for a review: Baserga, 1995).

Another crucial finding is that the transformed phenotype can be reversed by decreasing the number of IGF-IR using antisense strategies (Baserga, 1995; Resnicoff et al., 1994a; Resnicoff et al., 1994b; Shapiro et al., 1994; Neuenschwander et al. 1995; Resnicoff et al., 1995a; Resnicoff et al., 1995b; Lee et al., 1996; Pass et al., 1996; Baserga et al., 1997a) or by interfering with its function by expression of dominant negative mutants (Prager et al., 1994; D'Ambrosio et al., 1996).

The IGF-IR in cell survival

In vitro, an overexpressed IGF-IR protects cells from apoptosis induced by etoposide (Sell et al., 1995), tumor necrosis factor (Wu et al., 1996), IL-3-withdrawal (O'Connor et al., 1997; Prisco et al., 1997), p53 (Prisco et al, 1997), serum-deprivation (Parrizas et al., 1997), stress inducing agents in neuroblastoma cells (Singleton et al., 1996), okadaic acid (D'Ambrosio et al., 1997), ICE expression (Jung et al., 1996) and ionizing radiation (Kulik et al., 1997).

The protective effects of the IGF-IR are much more dramatic *in vivo*. A decrease in IGF-IR below normal levels (Resnicoff et al., 1995a; Resnicoff et al., 1995b) or expression of dominant negative mutant forms of the IGF-IR (D'Ambrosio et al., 1996) lead to induction of massive apoptosis of tumor cells *in vivo*.

Interestingly, the two major substrates of the IGF-IR (IRS-1 and Shc; Myers et al., 1993; White et al., 1994) do not protect cells from apoptosis as efficiently as the IGF-IR itself (Baserga et al., 1997b), suggesting that for cell survival the IGF-IR may have a pathway independent of IRS-1 and Shc.

Certain properties make the IGF-IR a potential and desirable target for therapeutic intervention:
1. it is ubiquitously expressed. Abnormalities in the expression of the IGF-IR and its ligands have been reported in a variety of tumors (Macaulay, 1992); they include upregulation and ectopic expression or overexpression (Arteaga, 1992; Kalebic et al., 1994);
2. targeting of the IGF-IR has modest growth inhibitory effect on cells in monolayer whereas it induces massive apoptosis under anchorage-independent conditions, such as those *in vivo* (Resnicoff et al., 1995a) . This property allows to discriminate between normal and tumor cells;
3. targeting of the IGF-IR in tumor cells results in massive apoptosis *in vivo*, leading to abrogation of tumorigenicity (Resnicoff et al., 1995b);
4. in addition to the apoptotic effect, targeting of the IGF-IR in tumor cells elicits an antitumor response in syngeneic immunocompetent animals leading to elimination of the residual cells escaping apoptosis. This CD8-dependent immune response protects the animals from subsequent tumor challenge (even when the challenge takes place more than 3 months later and at distant sites) and causes regression of established tumors with no recurrence observed during a year follow-up (Resnicoff et al., 1994a; Resnicoff et al., 1995b; Baserga et al., 1997b; Resnicoff et al., 1996).

In summary, the targeting of the IGF-IR acts a a double-edged sword inducing massive apoptosis of tumor cells *in vivo* and eliciting an antitumor response in the host that eliminates the few surviving cells and protects the animals from subsequent tumor challenge. These unique properties of the IGF-IR have been considered in the selection of the methods described below.

8.2 DETAILED PROTOCOLS

Sequences of antisense oligodeoxynucleotides (antisense-oligonucleotides) used to target the IGF-IR RNA

We use 18-mer phosphorothioate oligodeoxynucleotides (PS-oligonucleotides synthesized at Lynx Therapeutics, Hayward, Ca.); they are purified by HPLC and further processed to provide the sodium salt form of the final product.

We have tested several antisense sequences targeting different regions of the IGF-IR RNA; the most active compounds are the ones with sequences starting at the initiating methionine (CCG GAG CCA GAC TTC AT) or 3 (TCC TCC GGA GCC AGA CTT) or 6 nucleotides (GGA CCC TCC TCC GGA GCC) downstream of it. In addition, we have tested each sequence as fully thioate or end capped, with almost the same results. Antisense sequences to the middle region of the IGF-IR RNA (such as CTG CTC CTC CTC TAG GAT GA) are less active.

Controls (including sense, mismatch and random sequences) were tested in parallel.

Note: The synthetic oligodeoxynucleotides mentioned above contain CpG motifs, which have been reported to induce a B-cell-mediated immune response (for more information on this subject, please refer to Chapter 5). In our studies, we have also used cells expressing an antisense IGF-IR RNA (with no CpG motifs) with the same results, indicating that the observed effects are most likely due to the targeting of the IGF-IR rather than to artifacts.

Preparation of stock solutions

The PS-oligonucleotides are dissolved in aqueous medium; normally they come into solution very easily. Vortex can be used to allow complete dissolution.

The molecular weight of the 18-mer d(GGA CCC TCC TCC GGA GCC) phosphorothioate sodium salt is 5694 grams per mol. Thus, 5.694 mg of the above material (equivalent to 1 μmol) dissolved in 1 ml of medium will give a 1 mM stock solution.

If the sample is difficult to dissolve or if it changes the pH of the medium in which it is dissolved, this could be due to impurities. Nonspecific toxic effects can be expected under those conditions. Once the stock solution is prepared, the concentration is verified or adjusted by determining the OD at 260 nm. For the sequence mentioned above, the conversion factor provided by Lynx Therapeutics is 25.6 OD equal 1 mg of the above material. The oligonucleotides are then sterilized by filtration using a 0.22 μm pore-sized filter (Millipore, MA).

Lyophilized samples of PS-oligonucleotides can be stored at 4 °C. In order to avoid hydration, the tube or container should be wrapped in parafilm. Aqueous buffer solutions of PS-oligonucleotides (such as tris-EDTA, pH 8) have been shown to be stable for up to 3 months at either 4 or -20 °C.

Effects on the target. Determination of relative levels of IGF-IR

The effects of PS-oligonucleotides on IGF-IR expression at the protein level are determined by Western blot analysis.

1. Exponentially growing cells in monolayer are harvested by trypsinization and replated at 2.5×10^5 cells per 10 cm plate in appropriate serum-containing medium. The cells are then incubated at 37°C in an atmosphere of 5-7% CO_2.

2. Once the cells have attached and spread (this can take between 4 and 12 hours, depending on the cell line), the culture medium is removed by aspiration and the cells are washed three times with phosphate buffered saline (PBS). The cells are then shifted to the different conditions: serum-free medium supplemented with 0.1% bovine serum albumin fraction V (Sigma) and 1 μM ferrous sulfate (Sigma) alone or supplemented with 10 ng/ml of human recombinant IGF-I (Bachem, Ca) in the presence or absence of the different PS-oligonucleotides. Antisense and control sequences are tested in parallel. The doses of PS-oligonucleotides used for these studies ranged from 0.15 to 19 μM. The cells are incubated for 24 hours at 37°C.

3. The following day, the cells are washed with cold PBS and lysed in lysis buffer (50 mM HEPES pH 7.5; 150 mM NaCl; 1.5 mM $MgCl_2$; 1 mM EDTA; 10% glycerol; 1% Triton X-100 and protease inhibitors). Protein concentration is determined by Bradford assay. Equal amounts of protein are separated in 10% sodium-dodecyl-sulfate-polyacrylamide gel electrophoresis (SDS-PAGE) and transferred to nitrocellulose filters. After blocking with 5% non-fat milk in TBST (10 mM Tris-HCl pH7.5; 150 mM NaCl; 0.1% Tween 20), the filters are probed with a rabbit polyclonal antibody anti-β subunit of the IGF-IR (Santa Cruz Biotechnology, Ca.). The filters are carefully washed and blotted with a goat anti-rabbit antibody conjugated with horseradish peroxidase (Calbiochem). Detection is performed using the enhanced chemiluminiscence (ECL) reagent (Amersham). Relative levels of IGF-IR expression following treatment with PS-oligonucleotides are determined by scanning the film in a densitometer.

Autophosphorylation of the IGF-IR

As a functional assay for the IGF-IR, the autophosphorylation of the IGF-IR in response to IGF-I is analyzed. For this purpose, the cells are incubated in serum-free medium alone or supplemented with IGF-I in the presence or absence of the different oligonucleotides, as described above (steps 1 and 2).

The following day, the cells are lysed with the same lysis buffer described in step 3 and protein concentration is determined by Bradford. Equal amounts of protein are immunoprecipitated with a monoclonal antibody anti-IGF-IR (AB-1, Calbiochem) and protein A agarose (Calbiochem). The immunoprecipitates are separated in 10% SDS-PAGE and transferred to nitrocellulose filters. After blocking the filters with 5% non-fat milk in TBST, they are probed with a monoclonal antibody anti-phosphotyrosine conjugated with horseradish peroxidase (Transduction Laboratories). Detection is performed using the ECL reagent (Amersham).

In FO-1 human melanoma cells, a 90% decrease in IGF-IR levels is observed following treatment with antisense oligonucleotides (8 μM) whereas control oligonucleotides have no effect, even at 19 μM (Resnicoff et al., 1995a).

In CaOV-3 human ovarian carcinoma cells, pre-treatment with antisense oligonucleotides at 3 μM completely inhibits tyrosine phosphorylation of the IGF-IR in response to IGF-I whereas pre-treatment with control oligonucleotides does not have any effect (Resnicoff et al., 1993).

In C6 rat glioblastoma cells, pre-treatment with antisense oligonucleotides abolishes tyrosine phosphorylation of the β subunit of the IGF-IR in response to IGF-I stimulation (Resnicoff et al., 1994a). These effects correlate with a 50% decrease in the number of IGF-IRs, as determined by Scatchard analysis (Resnicoff et al., 1994a).

Pre-treatment with control oligonucleotides does not affect IGF-IR levels or its tyrosine phosphorylation (Resnicoff et al., 1994a).

Over the years we have used both Scatchard analysis and Western blotting to determine the effects of antisense oligonucleotides on the IGF-IR levels. We prefer the Western blot analysis because there is no interference with the IGF-I binding proteins.

Specificity of target

In order to verify that the antisense-oligonucleotides act specifically on IGF-IR expression, the same filter used for the determination of relative levels of IGF-IR (described above) is stripped with stripping buffer (100 mM β-mercaptoethanol; 2% SDS; 62.5 mM Tris HCl pH 6.7) at 50°C for 30 minutes, washed and reblotted using an antibody directed against another protein, such as epidermal growth factor receptor, focal adhesion kinase or Grb-2 just to name a few. The filter is carefully washed and blotted with a species-specific antibody conjugated with horseradish peroxidase. Detection is performed using the ECL reagent (Amersham). Relative levels of protein expression under the different conditions are determined by densitometry.

In vitro assays

Effects of antisense oligonucleotides to the IGF-IR RNA on IGF-I- dependent cell growth in monolayer

1. Exponentially growing cells are harvested by trypsinization and replated at a density of (2-4) x 10^4 cells per 35 mm. tissue culture plate in appropriate serum-containing medium. The cells are placed at 37°C in an atmosphere of 5-7% CO_2.
2. Once the cells have attached and spread (it can take between 4 and 12 hs depending on the cell line) the culture medium is removed by aspiration. The cells are carefully washed three times with PBS and shifted to the different experimental conditions: the appropriate serum-free medium (supplemented with 0.1% bovine serum albumin fraction V (Sigma) and 1 μM ferrous sulfate) alone or supplemented with 10 ng/ml of human recombinant IGF-I (Bachem, Ca.) in the presence or absence of the oligonucleotides. Antisense and control sequences are tested in parallel at a dose of 6.5 μM. The cells are incubated at 37 °C for 24 and 48 hours.
3. Cell number is determined at 24 and 48 hours following trypsinization and counting the cells in a hemocytometer. Cell viability is determined using the trypan blue exclusion assay.

C6 cells double in number after 24 hours and quadruplicate after 48 hours in culture in the presence of IGF-I. Incubation with antisense oligonucleotides at 6.5 μM completely abolishes the mitogenic response to IGF-I whereas control oligonucleotides have no effect on cell growth (Resnicoff et al., 1995a).

If 10 ng/ml of human recombinant IGF-I are not enough for optimal cell growth, the concentration of IGF-I can be increased up to 100 ng/ml

Effects of antisense oligonucleotides to the IGF-IR RNA on anchorage-independent growth: Colony formation in soft agar

1. 5×10^3 cells per 35 mm plate are seeded in Dulbecco's modified Eagle's medium (supplemented with 10% fetal bovine serum, penicillin, streptomycin and glutamin) containing 0.2% agarose with a 1% agarose underlay. Oligonucleotides at a dose of 9.5 µM are added to the medium prior to cell seeding. Antisense and control sequences are tested in parallel. Three plates are seeded per experimental condition.
2. The number of colonies and their size are determined after 3 weeks. Clonogenicity in soft agar is assayed by scoring the number of colonies larger than 125 µm.

C6 cells (untreated or treated with control oligonucleotides) have high clonogenic efficiency in soft agar (more than 100 colonies per 35 mm. plate) and form large colonies (> 250 µm in diameter).

C6 cells treated with antisense oligonucleotides at a dose of 9.5 µM have very low clonogenic efficiency in soft agar; they form less than 20 colonies per plate and in addition, the colonies formed are much smaller (Resnicoff et al., 1994a).

T98G human glioblastoma cells, untreated or treated with control oligonucleotides, have high clonogenic efficiency in soft agar (approximately 300 colonies are scored after 3 weeks). Following treatment with antisense oligonucleotides, as described above, less than 10 colonies are scored after 3 weeks after (Ambrose et al., 1994).

In vivo assays

Survival in vivo using the diffusion chamber assay

Materials needed for the construction of a diffusion chamber:
- 14 mm Lucite rings (two rings are required for each diffusion chamber: one ring with a loading hole and one ring without the hole; Millipore, MA.)
- 0.1 µm pore-sized Durapore membranes (2 membranes for each chamber; Millipore, MA)
- glue (Superglue™ or Krazyglue™)
- assembly tool designed to fit the membranes (Millipore, MA)

0.1 µm-pore sized Durapore membranes are chosen because they are hydrophilic and due to its pore size, they allow the passage of soluble factors (such as nutrients and proteins) excluding the exit or entry of intact cells. Thus, conditions inside the chambers reproduce the *in vivo* environment.

The assembly tool, made of Teflon and anodized aluminium, is required to align the membrane with the Lucite ring and to apply proper pressure while the glue dries. In addition, it avoids leakage due to imperfections during the assembly process and it guarantees that all the chambers are constructed in the same way.

Construction of the diffusion chambers

The chambers are constructed in two steps:
1. a membrane is glued to a Lucite ring, giving a half-chamber. Two types of half-chambers are constructed: one using rings with loading holes; another type is made using rings without holes. In order to make sure that all the parts are aligned, the membrane is placed in the assembly tool and a glued Lucite ring is placed on top. Pressure is applied using the upper part of the assembly tool while the glue dries overnight.
2. two half-chambers are glued giving a whole device. For this purpose, a half-chamber with a loading hole is glued to a half-chamber without the hole. Once again, several hours are required for the glue to dry. Prior to sterilization, it should be verified that the glue has not covered the loading hole in the ring; in such a case, the glue can be removed by passing a 23-gauge needle through the hole. This step is easier to do before sterilization of the chambers.

The whole device is then sterilized with ethylene oxide.

Bulldog clamps (Milltex Instrumental Company, New York) are used to hold the diffusion chambers.

Preparation of the cells for the diffusion chamber assay

1. Exponentially growing cells are harvested by trypsinization and replated at 2.5×10^5 cells per 10 cm plate in appropriate serum-containing medium. The cells are then incubated at 37 °C.
2. Once the cells have attached and spread (for C6 cells, it takes approximately 4 hours), the culture medium is removed by aspiration and the cells are washed three times with PBS. The cells are then shifted to the different experimental conditions: serum-free medium in the presence or absence of the different oligonucleotides (antisense and control sequences are tested in parallel). The doses of oligonucleotides used for these studies ranged from 0.15 to 19 µM. The cells are incubated for 24 hours at 37°C.
3. The following day, the cells are harvested by trypsinization and washed three times with PBS. The cells are counted using a hemocytometer and cell viability is determined by trypan blue exclusion. Cell viability should be more than 95% because antisense oligonucleotides to the IGF-IR RNA do not induce cell death under anchorage-dependent conditions, such as monolayer cultures.

4. 5×10^5 cells suspended in 0.2 ml PBS are loaded in each chamber. The cells are loaded by injection through the loading hole of the chamber using a 23 gauge needle. Once the cells have been injected into the chamber, the hole is plugged with a plastic thread provided by Millipore. All these procedures are done under sterile conditions. Each experimental condition is tested in triplicate.

The diffusion chambers are implanted in the subcutaneous tissue of 7 week-old rodents under anesthesia with isofluorine, which is an inhalant that allows fast recovery. The incision wound is suturated with staples.

At the desired intervals, the diffusion chambers are removed from the animals, under anesthesia, and the cells are recovered under sterile conditions.

Recovery of cells from the diffusion chambers

1. The outside of the chamber is cleaned with a cotton swab to eliminate exudate from the animals.
2. A small incision is made on the upper membrane and through this hole, the medium is aspirated with a pipette and transferred to an Eppendorf tube.

The volume recovered from the chamber should be the same one as loaded (0.2 ml); otherwise, it could be due to leakage.

The cells do not attach to the membrane but they sediment and deposit at the bottom of the chambers. For this reason, it is very important to keep the chambers in the same position they have been recovered from the animals. The cells could be lost if the incision is made in the membrane where the cells have sedimented.

3. A bigger incision is then made on the upper membrane to allow thorough washes of the chamber with PBS, using a pipette and making sure to recover all the cells.
4. The cells transferred to an Eppendorf tube are then washed three times with PBS and counted using a hemocytometer. Cell viability is assessed by trypan blue exclusion. Results are expressed as percentage of viable cells recovered from the original inoculum.

Tumor cells implanted in the diffusion chambers double in number after 24 hours *in vivo*, yielding a recovery of 200% (or more) of the original inoculum. Cells pre-treated with antisense oligonucleotides and recovered after 24 hours *in vivo*, undergo massive apoptosis; cell recovery is less than 100%, depending on the dose of antisense oligonucleotide used and the cell line tested (Resnicoff et al., 1995a; Resnicoff et al., 1995b; Baserga et al., 1997b).

C6 cells untreated or pre-treated with control oligonucleotides (even at doses as high as 19 µM for 24 hours prior to *in vivo* implantation) give a recovery of 200% of the original inoculum, demonstrating that the diffusion chambers allow optimal cell growth *in vivo* (Resnicoff et al., 1995a; Resnicoff et al., 1995b).

C6 cells pre-treated with antisense oligonucleotide GGA CCC TCC TCC GGA GCC at a dose as low as 0.15 μM for 24 hours prior to *in vivo* implantation of the cells, give a recovery of 54% of the original cell number. This means that 46% of the implanted cells have died within 24 hours *in vivo*. If the dose of antisense oligonucleotide is increased to 16 μM, cell recovery is 0.01% of original inoculum, meaning that 99.99% of the implanted cells have died within 24 hours *in vivo* (Resnicoff et al., 1995a; Resnicoff et al., 1995b).

This same sequence of antisense oligonucleotide have been used for the tumorigenesis assays.

This same protocol has been used to target the IGF-IR in primary cultures of human brain tumors (glioblastomas multiforme grade IV and astrocytomas grade II). Following treatment with antisense oligonucleotide (9.5 μM for 24 hours), cell recovery after 24 hours *in vivo* ranges from 0.3 to 1.5% of the original inoculum, depending on the tumor. Cultures pre-treated with control oligonucleotides give the same recovery as untreated cells.

Tumorigenesis in nude mice

1. The cells are pre-treated with the oligonucleotides and prepared for injection as described above for the diffusion chamber assay (steps 1-3).
2. For C6 cells, 10^5 cells suspended in 0.1 ml of PBS are injected subcutaneously (s.c.) above the hind leg of 7-week-old male Balb/c nude mice (Charles River Breeders). A minimum of three mice are injected per experimental condition.

The injected volume should not exceed 0.1 ml because the skin of the nude mice cannot absorb more at once; otherwise, a drop will come out and there will be more variability among the animals of each group. Males are preferred over females to avoid hormonal variation.

3. The animals are monitored for tumor development on a daily basis.

For C6 cells, untreated or pre-treated with control oligonucleotides even at high doses, tumors become palpable within 4 days of injection. Tumors grow fast and become bulky after a month; at that point, the animals need to be sacrificed to comply with institutional guidelines for animal welfare (Resnicoff et al., 1995b).

C6 cells pre-treated with antisense oligonucleotides can develop tumors in nude mice, but the tumors appear with a delay proportional to the number of cells escaping apoptosis. For example, C6 cells pre-treated with antisense oligonucleotides at 1.5 μM give a cell recovery of 35%;

considering the number of cells escaping apoptosis and assuming a cell doubling time *in vivo* of 24 hours, tumors should become palpable by day 7. In fact, tumors develop after 11 days. This discrepancy between the estimated time of tumor appearance and the real time at which tumors become palpable could be due to the following: a) we have underestimated the extent of apoptosis *in vivo* by measuring it only after 24 hours; eventually, more cells could die thereafter; b) we have overestimated the cell doubling time *in vivo*; it could be that when very few cells survive, it takes more than 24 hours for the cells to double *in vivo*. At any rate, the correlation between the number of cells escaping apoptosis *in vivo* and the time of tumor appearance in nude mice is still valid since we have never observed tumors appearing before the estimated time (Resnicoff et al., 1995b).

Tumorigenesis in syngeneic animals

Two syngeneic models have been used:
C6 rat glioblastoma cells in BD IX rats
B1792-F10 mouse melanoma in C57/BL6 mice

C6 rat glioblastoma cells in BD IX rats

C6 cells are pre-treated with oligonucleotides and prepared for injection as described above for the diffusion chamber assay (steps 1-3).

10^7 cells suspended in 0.2 ml PBS are injected s.c. above the hind leg of 7-week-old male BD IX rats (Charles River Breeders). A minimum of three rats are injected per experimental condition.

The rats are monitored for tumor development on a daily basis.

C6 cells untreated or pre-treated with control oligonucleotides (even at high doses, such as 19 µM), develop palpable tumors within 4 days of injection; tumors progress and by a month, the animals need to be sacrificed upon the development of bulky tumors in order to comply with institutional guidelines for animal welfare (Resnicoff et al., 1995b).

For C6 cells pre-treated with antisense oligonucleotides at 3 µM (or higher doses), no tumors develop after more than 4 months of injection (Resnicoff et al., 1995b). For C6 cells pre-treated with antisense oligonucleotides between 0.15 and 2.5 µM, tumors appear by day 5; they become smaller by day 19 and they completely regress by day 25 with no recurrence observed during more than 4 months follow-up (Resnicoff et al., 1995b). All of these rats are fully protected against subsequent tumor challenge, even when it takes place more than 3 months later and at distant sites (Resnicoff et al. 1995b).

B1792-F10 mouse melanoma cells in C57/BL6 mice

The cells are pre-treated with oligonucleotides and prepared for injection as described above for the diffusion chamber assay (steps 1-3).

10^5 cells suspended in 0.2 ml PBS are injected s.c. above the hind leg of 7-week-old male C57/BL6 mice. A minimum of three mice are injected per experimental condition.

The mice are monitored daily for tumor development.

B1792-F10 cells (untreated or pre-treated with control oligonucleotides even at 19 μM) develop palpable tumors within 4 days of injection; animals injected with tumor cells pre-treated with antisense oligonucleotides at doses between 13 and 19 μM remain tumor-free for more than 3 months (Baserga et al., 1997b).

The mice that received the antisense-treated cells are protected against subsequent tumor challenge; full protection is conferred by tumor cells pre-treated with antisense oligonucleotides at a dose of 19 μM whereas partial protection is observed for a dose of 13 μM (tumors appeared after 12 days; Baserga et al., 1997b). Mice injected with untreated cells or pre-treated with control oligonucleotides, even at 19 μM, are not protected against subsequent tumor challenge and develop bilateral tumors within 5 days (Baserga et al., 1997b).

8.3 CONCLUDING REMARKS

Given that the IGF-IR is not crucial for cell growth in monolayer, it should not surprise to observe modest growth inhibitory effects following treatment of tumor cells with antisense oligonucleotides to the IGF-IR RNA in monolayer cultures.

For a given sequence of antisense oligonucleotide, a minimum dose of 9.5 μM is required to cause 50% inhibition of IGF-I-dependent cell growth in monolayer whereas a dose of 0.15 μM is able to kill 67% of the cells in the diffusion chamber assay in vivo.

The diffusion chamber assay *in vivo* is the most sensitive of all the assays mentioned above.

It has allowed us to discriminate among different antisense oligonucleotides which were all equally active in *in vitro* assays; in addition, we were able to select the most active compound, screening for activity at doses as low as 0.15 μM.

For these reasons, we have routinely screened our compounds for activity using the diffusion chamber assay *in vivo*, which is very sensitive and very fast, giving *in vivo* results in 24 hours.

8.4 SUMMARY

This chapter describes the methods designed to test for insulin-like growth factor I receptor (IGF-IR) function in tumor cells following treatment with antisense oligodeoxynucleotides to the IGF-IR RNA.

These methods were chosen based on certain properties of the IGF-IR:
a) the IGF-IR is ubiquitously expressed. Abnormalities in the expression of the IGF-IR and its ligands have been reported in a variety of tumors; b) the IGF-IR is not crucial for cell growth but it is essential for the establishment and maintenance of the transformed phenotype and for cell survival in several cell types; c) targeting of the IGF-IR has a modest growth inhibitory effect on cells in monolayer but it induces massive apoptosis under anchorage-independent conditions, such as those *in vivo*. Thus, targeting of the IGF-IR has a discriminatory effect between normal and tumor cells; d) targeting of the IGF-IR in tumor cells results in massive apoptosis, leading to abrogation of tumorigenesis; e) in addition to the apoptotic effect, targeting of the IGF-IR in tumor cells elicits an antitumor response in syngeneic immunocompetent animals, leading to elimination of the residual cells escaping apoptosis.

This antitumor response protects the animals from subsequent tumor challenge and it causes complete regression of established tumors.

Altogether, these properties make the IGF-IR a desirable target for therapeutic intervention.

8.5 REFERENCES

Ambrose D, Resnicoff M, Coppola D, Sell C, Miura M, Jameson S, Baserga R, Rubin R. Growth regulation of human glioblastoma T98G cells by insulin-like growth factor-1 and its receptor. *J Cell Physiol* 1994;159:92-100

Arteaga CL. Interference of the IGF system as a strategy to inhibit breast cancer growth. *Breast Cancer Res Treat* 1992;22:101-06

Baker J, Liu J-P, Robertson EJ, Efstratiadis A. Role of insulin-like growth factors in embryonic and postnatal growth. *Cell* 1993;75:73-82

Baserga R. The insulin-like growth factor 1 receptor: a key to tumor growth? *Cancer Res* 1995;55:249-52

Baserga R, Resnicoff M, D'Ambrosio C, Valentinis B. The role of the IGF-1 receptor in apoptosis. *Vitamins & Hormones* 1997a; 53:65-98

Baserga R, Resnicoff M, Dews M. The IGF-IR and Cancer. *Endocrine* 1997b;7:99-102

D'Ambrosio C, Ferber A, Resnicoff M, Baserga R. A soluble insulin-like growth factor 1 receptor that induces apoptosis of tumor cells *in vivo* and inhibits tumorigenesis. *Cancer Res* 1996;56:4013-20

D'Ambrosio C, Valentinis B, Prisco M, Reiss K, Rubini M, Baserga R. Protective effect of the Insulin-like growth factor 1 receptor on apoptosis induced by okadaic acid. *Cancer Res* 1997;57:3264-71

Jung Y-K, Miura M and Yuan Y. Suppression of interleukin-1β converting enzyme-mediated cell death by insulin-like growth factor. *J Biol Chem* 1996;271:5112-17

Kalebic T, Tsokos M, Helman LJ. In vivo treatment with antibody against IGF-1 receptor suppresses growth of human rhabdomyosarcoma and downregulates p34 cdc2. *Cancer Res* 1994;54:5531-34

Kulik G, Klippel A, Weber MJ. Antiapoptotic signaling by the insulin-like growth factor 1 receptor, phosphatidylinositol 3-kinase and akt. *Mol Cell Biol* 1997;17:1595-606

Lee CT, Wu S, Gabrilovich D, Chen H, Nadaf-Rahrov S, Ciernick IF, Carbone DP. Antitumor effect of an adenovirus expressing antisense insulin-like growth factor I receptor on human lung cancer cell lines. *Cancer Res* 1996;56:3038-41

Liu J-P, Baker J, Perkins antisense, Robertson EJ, Efstratiadis A. Mice carrying null mutations of the genes encoding insulin-like growth factor 1 (IGF-1) and type 1 IGF receptor (igf1r). *Cell* 1993;75:59-72

Macaulay VM. Insulin-like growth factors and cancer. *Brit J Cancer* 1992;65:311-20

Myers MG, Sun XJ, Cheatham B, Jachna BR, Glasheen EM, Backer JM, White MF. IRS-1 is a common element in insulin and insulin-like growth factor 1 signaling to the phosphatidylinositol 3' kinase. *Endocrinology* 1993;132:1421-30

Neuenschwander S, Roberts CT Jr, LeRoith D. Growth inhibition of MCF-7 breast cancer cells by stable expression of an insulin-like growth factor 1 receptor antisense ribonucleic acid. *Endocrinology* 1995;136:4298-303

O'Connor R, Kauffmann-Zeh A, Liu Y, Lehar S, Evan GI, Baserga R, Blattler WA. The IGF-1 receptor domains for protection from apoptosis are distinct from those required for proliferation and transformation. *Mol Cell Biol* 1997;17:427-35

Parrizas M, Saltiel A, LeRoith D. Insulin-like growth factor 1 inhibits apoptosis using the phosphatidylinositol 3' kinase and mitogen-activated protein kinase pathways. *J Biol Chem* 1997;272:154-61

Pass HL, Mew DJY, Carbone M, Matthews WA, Domington JS, Baserga R, Walker CI, Resnicoff M, Steinberg SM. Inhibition of hamster mesothelioma tumorigenesis by an antisense expression plasmid to the insulin-like growth factor 1 receptor. *Cancer Res* 1996;56:4044-48

Prager D, Li HL, Asa S, Melmed S. Dominant negative inhibition of tumorigensis *in vivo* by human insulin-like growth factor receptor mutant. *Proc Natl Acad Sci USA* 1994;91:2181-85

Prisco M, Hongo A, Rizzo MG, Sacchi A, Baserga R. The IGF-1 receptor as a physiological relevant target of p53 in apoptosis caused by interleukin-3-withdrawal. *Mol Cell Biol* 1997;17:1084-92

Resnicoff M, Ambrose D, Coppola D, Rubin R. Insulin-like growth factor-1 and its receptor mediate the autocrine proliferation of human ovarian carcinoma cell lines. *Lab Invest* 1993;69:756-60

Resnicoff M, Sell C, Rubini M, Coppola D, Ambrose D, Baserga R, Rubin R. Rat glioblastoma cells expressing an antisense RNA to the insulin-like growth factor 1 (IGF-1) receptor are nontumorigenic and induce regression of wild-type tumors. *Cancer Res* 1994a;54:2218-22

Resnicoff M, Coppola D, Sell C, Rubin R, Ferrone S, Baserga R. Growth inhibition of human melanoma cells in nude mice by antisense strategies to the type 1 insulin-like growth factor receptor. *Cancer Res* 1994b;54:4848-50

Resnicoff M, Abraham D, Yutanawiboonchai W, Rotman HL, Kajstura J, Rubin R, Zoltic P, Baserga R. The insulin-like growth factor 1 receptor protects tumor cells from apoptosis *in vivo*. *Cancer Res* 1995a;55:2463-69

Resnicoff M, Burgaud J-L, Rotman HL, Abraham D, Baserga R. Correlation between apoptosis, tumorigenesis and levels of insulin-like growth factor 1 receptors. *Cancer Res* 1995b;55:3739-41

Resnicoff M, Tjuvajev J, Rotman HL, Abraham D, Curtis M, Aiken R, Baserga R. Regression of C6 rat brain tumors by cells expressing an antisense insulin-like growth factor 1 receptor RNA. *J Exp Ther Oncol* 1996;1:385-89

Sell C, Rubini M, Rubin R, Liu J-P, Efstratidis A, Baserga R. Simian virus 40 large tumor antigen is unable to transform mouse embryonic fibroblasts lacking type 1 IGF receptor. *Proc Natl Acad Sci USA* 1993;90:11217-21

Sell C, Dumenil G, Deveaud C, Miura M, Coppola D, De Angelis T, Rubin R, Efstratiadis A, Baserga R. Effect of a null mutation of the type 1 Igf receptor gene on growth and transformation of mouse embryo fibroblasts. *Mol Cell Biol* 1994;14:3604-12

Sell C, Baserga R, Rubin R. Insulin-like growth factor 1 (IGF-1) and the IGF-1 receptor prevent etoposide-induced apoptosis. *Cancer Res* 1995;55:303-06

Shapiro DN, Jones BG, Shapiro LH, Dias P, Houghton PJ. Antisense-mediated reduction in insulin-like growth factor 1 receptor expression suppresses the malignant phenotype of a human alveolar rhabdomyosarcoma. *J Clin Invest* 1994;94:1235-42

Singleton JR, Randolph AE, Feldman EL. Insulin-like growth factor 1 receptor prevents apoptosis and enhances neuroblastoma tumorigenesis. *Cancer Res* 1996;56:4522-29

Ullrich A, Gray A, Tam AW, Yang-Feng T, Tsubokawa M, Collins C, Henzel W, LeBon T, Kahuria S, Chen E, Jakobs S, Francke U, Ramachandran J, Fujita-Yamaguchi Y. Insulin-like growth factor 1 receptor primary structure: comparison with insulin receptor suggests structural determinants that define functional specificity. *EMBO J* 1986;5:2503-12

White MF, Kahn CR. The insulin signaling system. *J Biol Chem* 1994;269:1-4

Wu Y, Tewari M, Cui S, Rubin R. Activation of the insulin-like growth factor 1 receptor inhibits tumor necrosis factor-induced cell death. *J Cell Physiol* 1996;168:499-509

Part III:
Antisense Application
in vivo

9 How to Test Antisense Oligonucleotides in Animals

Rainer Spanagel[1], Christoph Probst[2], Deborah C. Mash[3] and T. Skutella[4]

[1]Max Planck Institute of Psychiatry,
Drug Abuse Group,
Munich, Germany
[2]Max Planck Institute of Neurobiology
Martinsried, Germany
[3]School of Medicine
Miami, Florida, USA
[4]Humboldt University
Institute of Anatomy
Berlin, Germany

9.1 INTRODUCTION

As their name implies, antisense oligonucleotides (oligonucleotides) are thought to bind with mRNA molecules in a complementary fashion and either physically block protein biosynthesis or activate endogenous nucleic acid degrading enzymes. The rational behind the *in vivo* antisense approach is to down-regulate the expression of a specific gene in order to evaluate the potential biological function of the corresponding protein. One of the particular major advantages offered by the antisense approach is the possibility to focus on neurochemical systems that, because of the lack of selective neurochemical tools, have so far escaped detailed functional analysis. The results of *in vivo* antisense targeting experiments in the past have led to mixed feelings regarding this technology, ranging from amazement and satisfaction to frustration. Because *in vivo* oligonucleotide technology is not void of methodological problems that can affect the validity and selectivity this approach might sometimes lead to misinterpretations when working with yet unexplored neurochemical systems.

The first crucial point to be considered when planning antisense targeting experiments is the chemistry or biochemistry of the synthetic nucleic acids (nucleotide sequence and chemical modification) when introduced into the living organism. The next step is dosage as well as the route and duration of administration. Furthermore, specificity of the antisense approach has to be warranted by the appropriate controls. What follows is a review of the chemical modifications of oligonucleotides and guidelines regarding nucleotide sequence and oligonucleotide length.

In order to validate the antisense approach in more detail we manipulated specific brain systems which had already been the subject of extensive pharmacological investigations. Both the mesolimbic and nigrostriatal dopamine (DA) system provide an ideal model system because they have previously been analyzed with various neurochemical tools and the role of endogenous DA has been studied intensively with experimental depletion using specific neurotoxins (such as 6-OHDA, MPTP; for review see Reading and Dunnett, 1994) or specific DA-receptor antagonists. These effects have been thoroughly characterized on the neurochemical and behavioral level. Using these well characterized brain systems, in a first set of experiments we analyzed the neurochemical and behavioral effects of antisense oligonucleotides aimed at tyrosine hyroxylase (TH) mRNA, the transcriptional message of the rate limiting enzyme in the metabolic pathway leading to catecholamine synthesis. In a second set of experiments we studied the functional role of various DA-receptors by *in vivo* antisense targeting. These two examples of targeting an enzyme system (tyrosin hydroxylase TH) and a receptor system (DA-receptors) will be described in greater depth in order to provide detailed methodological protocols.

9.2 CHEMICAL MODIFICATIONS OF OLIGONUCLEOTIDES

Here we will give an overview over the most common chemical modifications being introduced into oligonucleotides that are used in *in vivo* experiments. The objectives for generating modified oligonucleotides are primarily the protection of the reagents from nucleolytic attack and degradation, increased binding affinity and cellular uptake. Since oligonucleotides are highly charged molecules, transmembrane transport will not be a favored process in the biophysical view. Efforts to improve the biophysical properties of antisense oligonucleotides have been made by modifying the phosphate backbone of the oligonucleotide (figure 1D) or by incorporating sugar modified nucleoside analogs (figure 1A). Another approach uses a combination of different modifications and the resulting chimeric antisense oligonucleotides show increased nuclease resistance (figure 1B), improved cellular uptake (figure 1C) and a higher binding affinity to the target mRNA (figure 1E).

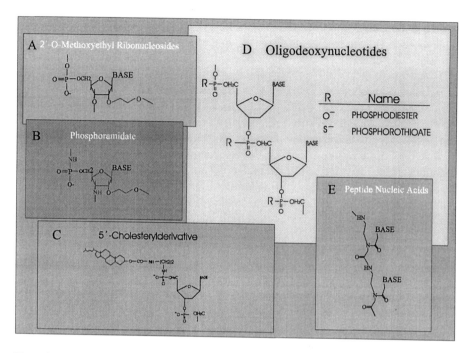

Figure 1. Various chemical modifications of oligonucleotides and other constructs

The precise mechanisms of antisense action are so far unknown. RNAase H activation and translational arrest seem to be the most prominent molecular events resulting from oligonucleotide binding (Skutella et al., 1994a; Probst and Skutella, 1996a; Baker et al., 1997). The antisense oligonucleotides seem to inhibit the assembly of the translational complex consisting of initiation factors and ribosome subunits. This blocks the *de novo* biosynthesis of proteins and the resulting protein deficit leads to a loss of function. However, as antisense experiments have shown, application of antisense oligonucleotides do not lead to a complete loss of protein. They down-regulate protein expression approximately 40-80%, which usually results in sufficient protein deficiencies to lead to observable effects. Nevertheless, complete mRNA depletion would be favorable. This is the reason for searching for modifications that activate endogenous RNA degrading enzymes or for designing antisense oligonucleotides which incorporate enzymatic activity. Replacement of one oxygen atom in the phosphoric acid backbone of the nucleic acid molecules with sulfur results in a relatively stable and nuclease resistant derivative (figure 1D). The resulting phosphorothioate oligonucleotides are by now the most widely used antisense derivatives in antisense targeting experiments. These oligonucleotides have proven to be quite effective, as documented in a large number of publications although there are still several negative aspects to be considered. The introduction of non-physiologic sulfur building blocks seems to impair the normal cellular machinery which may lead to unwanted side effects in *in vivo* experiments (Spanagel et al., 1998). A reduction of the sulfur content by using so-called end-capped oligonucleotides resulted in a marked decrease in the side effect spectrum without affecting the longevity and efficiency of the antisense oligonucleotides *in vivo* (Skutella et al., 1994b; Schmidt et al. 1997). This approach has now been further developed in the form of minimally modified oligonucleotides or a combination of end-capping and pyrimidine protection (Peyman and Uhlmann, 1996; Hebb and Robertson, 1997). Another reason why phosphorothioate modification seems to be so widely distributed is its ease of synthesis and relatively low costs. High quality phosphorothioate oligonucleotides can be ordered by almost every commercial DNA-supplier in sufficient quantities. Alternatively, phosphoramidate modified oligonucleotides (figure 1B) have a higher affinity to single stranded RNA and are therefore more potent than conventional phosphorothioate oligonucleotides (Skorski et al., 1997). Additionally it has been shown that the activities of the phosphoramidate oligonucleotides are not depending on nuclease RNAse H activity (Heidenreich et al., 1997).

Other modifications require more sophisticated synthesis strategies that are not commercially available or are still extremely expensive. In order to direct oligonucleotides selectively to discrete cellular compartments, they can be coupled to peptide moieties. If the sequence of the peptide corresponds to specific signal sequences the oligonucleotide may be directly routed according to the peptide routing signal (Arar et al., 1995). For example nuclear targeting peptide-antisense oligonucleotides showed increased nuclear uptake and antisense activity in a model system using the freshwater ciliate *Paramecium* (Reed et al., 1995). Antisense oligonucleotides may not only bind to respective mRNA molecules. An additional target in the cell is found in the nucleus, the genomic DNA. Here polypyrimidine oligonucleotides may bind in the major groove of the DNA double-helix, producing a triple-helix. Triple-helix-producing oligonucleotides may be more effective than mRNA binding oligonucleotides because they block the *de novo* synthesis of RNA (Mergny et al., 1992) but triple-helix oligonucleotides are still in an investigational status and are yet not applied in animal studies even though first cell culture experiments look very promising (Aggrawal et al., 1996). The same is true for another novel modification strategy which represents a completely different type of oligonucleotide modification. Here, the complete deoxyribose backbone is replaced with a backbone similar to that found in amino acids (Nielsen et al., 1996). The polyamides or peptide nucleic acids (figure 1E) show a high degree of nuclease and protease resistance combined with greater specificity and tighter binding properties when compared with conventional DNA/DNA, RNA/RNA or DNA/RNA hybrids. The problems associated with *in vivo* bioefficacy of the peptide nucleic acids are, however, still unsolved and this has so far hampered *in vivo* application of these oligonucleotide derivatives (Basu and Wickstrom, 1997). As already mentioned, there are activities towards developing antisense oligonucleotides with the intrinsic property of nucleic acid scission. 2´, 5´-Oligoadenylate antisense chimeras utilize the specificity of the antisense concept combined with the potent RNA-degrading activity of the 2-5A-dependent endonuclease RNAse L. The trimeric oligonucleotide ppp5´A2´p5´A2´p5´A activates RNAse L that is part of the 2-5A system and seems to mediate certain interferon actions (Torrence et al., 1993; Cirino et al., 1997). So far only cell culture data is available demonstrating the effectiveness of the 2-5A-oligonucleotides.

Another promising approach is the use of catalytic active RNA molecules so called ribozymes. The ribozymes are recycled and may cleave many target RNA molecules. Antisense application is proceeding very quickly, stimulated by the fact that naturally occurring ribozymes act very efficiently

Taken together, in the last few years, the development of antisense oligonucleotides has made substantial progress. The chemistry of novel modifications has led to oligonucleotides with improved properties such as enhanced cellular uptake, increased potency and decreased toxic side effects, which will broaden the scope of their application.

9.3 OLIGONUCLEOTIDE SEQUENCE AND LENGTH

When considering oligonucleotide sequence and length, the selectivity of hybridization and the cellular uptake probability require consideration. Cellular uptake should pose a problem, especially when taking a closer look at the oligonucleotide chemistry. The highly charged nucleic acid molecules do not cross the lipid cell membrane readily unless there is a transport mechanism. One such possible mediator mechanism is binding to Mac-1 (Stein et al., 1997). Fact is, that the oligonucleotides are incorporated by the cell and are found both in the cytoplasm and nucleus. In cell culture experiments, uptake was shown to be rapid and stable (Li et al., 1997; Skutella et al., 1998) and recent results do show that internalization of oligonucleotides may follow "binding protein"-mediated endocytotic mechanisms (Beck et al., 1996). Cellular uptake is size dependent where smaller oligonucleotides are taken up by the cell more efficiently (Loke et al., 1989).

Oligonucleotide cellular uptake efficiency is not the only determinant that has to be taken into account when designing the antisense probe. In determining the optimal oligonucleotide size one has to consider the size of the genome and the probability of finding a given sequence in the genome. The chance of finding the sequence in the four letter genetic alphabet is $1:4^x$ with x denoting the number of bases which are incorporated in an oligonucleotide. Therefore, a 4mer oligonucleotide will have a chance of 1 in $4^4 = 256$ different sequences or be unique in a genome of the size of 256 base pairs. A typical mammalian genome has a size of approx. 10^9 base pairs. A 18mer oligonucleotide will have a chance of repetition in 1 in 6.9×10^{10} bases, i.e. almost no chance of existing twice in the genome, whereas smaller oligonucleotides, such as an 8mer will be found one in every 65536 bases. This is based on the assumption that each of the four bases have equal distribution probabilities. Reality is a little bit different. Due to the fact that many proteins, e.g. larger multidomain proteins, evolved through a process termed exon shuffling, a number of conserved sequence motifs will be found in various homologous proteins comprising protein families. This needs to be taken into consideration when designing oligonucleotides and analyzing the database comparisons. In conclusion, a critical analysis of the database searches has to be performed in order to correctly evaluate and interpret the search results.

Perfect homology is still no guarantee of full specificity as it is impossible to control oligonucleotide hybridization kinetics in the *in vivo* situation. So, with the size of 18 base pairs a compromise was found experimentally where an optimum in specificity and cellular uptake was found. A plethora of publications prove the effectiveness of this approach. (Monia 1996; Monia et al., 1997; Moulds et al., 1995) In the end antisense construction has to be followed by experimental optimization, in general by trial and error to find the most effective antisense agent. Potential paradigms for antisense design have usually not proven to be valid and the application of combinatorial techniques to screen large numbers of oligonucleotides is not feasible in normal laboratory practice.

9.4 OLIGONUCLEOTIDE SPECIFICITY AND APPROPRIATE CONTROLS FOR *IN VIVO* EXPERIMENTS

Specificity is a feature strictly related to the experimental conditions and, for this reason, has to be established afresh each time, even if the same oligonucleotide has already been used repeatedly. Specificity can also be overshadowed by the induction of sequence independent non-specific effects as well as of a few short-term effects e.g. electrophysiological neuronal responses that are sequence specific but target independent (Neumann et al., 1995). The problems associated with non-specific effects have recently been reviewed extensively (Spanagel et al. 1998). In brief, intracerebroventricular (i.c.v.) injections/infusions of oligonucleotides are accompanied by fever and sickness-like behavior (i.e. behavioral depression, anorexia and adipsia) (Schöbitz et al., 1997). These symptoms occur in response to various antisense, sense and missense oligonucleotides and are not affected by the chemical modification of these molecules. We also recently observed that non-specific behavioral and autonomic changes in response to centrally administered oligonucleotides are associated with induction of interleukin-6 in the brain, indicating that nucleic acids are endowed with proinflammatory properties (Schöbitz et al., 1997). In conclusion, the administration of oligonucleotides into the lateral ventricle produce a number of non-specific effects that warrant caution when the effects of the suppression of a particular protein are examined with the antisense technology. For instance, it is questionable whether experiments on body temperature control involving i.c.v.-injected oligonucleotides can yield valid data. In addition, we observed that ongoing acquired behavior (food-reinforced behavior) is non-specifically reduced after i.c.v. injection of missense oligonucleotides, but that this effect can be prevented by local injection into the brain parenchyma (Skutella et al., 1994c; Spanagel et al., 1998). However, other experiments showed that i.c.v. administered missense oligonucleotides do not alter spontaneous ongoing or stimulated behavior determined in the elevated plus maze or in response to exposure to a conspecific juvenile (Spanagel et al., 1998). We therefore conclude that oligonucleotides do not necessarily influence the results of behavioral paradigms in a non-specific manner if (i) appropriate controls are used, including both missmatch-base oligonucleotides and vehicle (see also guidelines by Stein and Krieg, 1994), (ii) non-specific effects of oligonucleotides on body temperature and general activity under resting conditions are taken into account and (iii) the dosage is reduced or the oligonucleotide is administered locally. Finally it is unavoidable to measure as many parameters as possible e.g. effects on mRNA and protein levels, behavioral effects due to protein deficits or other secondary effects resulting from the loss of gene function (see 9.5.3.).

9.5 MODE AND TIME-WINDOW OF *IN VIVO* ADMINISTRATION OF OLIGONUCLEOTIDES

Various time courses of oligonucleotide application have been described so far. As with the problems described in the preceding section, optimal time courses and concentrations are hampered by insufficient knowledge of the actual *in vivo* cellular conditions, though there are a few reports comparing single acute, repeated intermittent and continous infusions of oligonucleotides. The results of numerous studies using either acute or continous infusions are reviewed by Landgraf and coworkers (1996; 1997). In general, the length of antisense treatment depends on the turnover of the targeted peptide or protein. Wherever possible, thorough knowledge of the mRNA and protein turnover rates would be beneficial and help to choose appropriate duration of antisense treatment. Time dependence has been described in many reports and we have also observed it in our own experiments, so that we believe that chronic treatment by continuous infusion via osmotic minipumps is the method of choice to produce an adequate downregulation of the target gene (figure 2). Even though short term antisense treatments are also reported with effective inhibitory effects, e.g acute i.c.v. injection of arginine vasopressin antisense oligonucleotides not only significantly reduced arginine vasopressin immunoreactivity in the hypothalamus but also affected behavioral parameters like licking rate 6 h after antisense treatment (Skutella et al., 1994a). Another difficulty is encountered when proteins are sequestered in pools so that antisense treatment may hinder *de novo* synthesis of the corresponding mRNA, but due to the spare protein, no physiological effects will be observed. The same will be true for various receptor molecules in that even if 90 % of the mRNA were depleted, the remaining 10 % would suffice to fully maintain second messenger activation (Standifer et al., 1994).

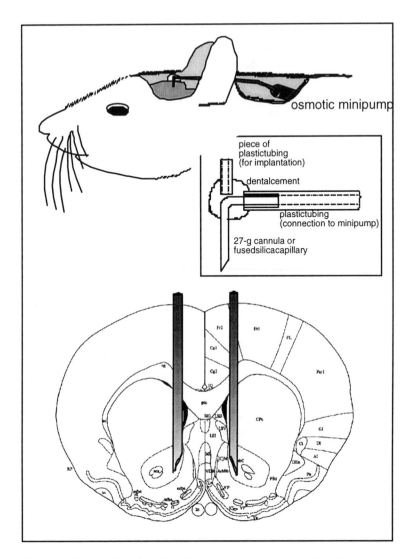

Figure 2. Bilateral oligonucleotide infusion into the nucleus accumbens shell region via osmotic mini-pumps. The entire device, i.e. the infusion cannula already connected to the plastic tubing and the ALZET pump, is implanted. The whole infusion device is filled with either the antisense or a control solution. We usually use a relatively long piece of plastic tubing and insert a very small air bubble near the pump to control the rate of delivery *post mortem*.

9.6 *IN VIVO* ANTISENSE TARGETING OF THE ENZYME TYROSIN HYDROXYLASE (TH)

In the experiments described here we selected and determined optimal sequences for oligonucleotides by using computer programs to test sequence similarity and binding (Ho et al., 1996). Binding was simulated for each selected oligonucleotide against all known gene sequences in specified databases such as GenBank and EMBL. To establish efficient antisense oligonucleotides it might be reasonable to test selected oligonucleotides for potency and specificity in simplistic environments first, such as cell culture systems. The selected oligonucleotides might then be extrapolated to more complex tasks such as *in vivo* models. In our experiments we used unmodified and phosphorodiester, 3` or 3`-5` capped or complete phosphorothioate modifications. The sequence of the 20-base antisense oligonucleotide used in the present study was designed according to the primary structure of the mRNA that encodes TH. The effective antisense sequence (GGT GGG CAT AGT GCA AGC TG) corresponded to the NH_2-terminus around the initiating codon. Mismatch sequence oligonucleotide (GAT CCG CAT AGG GCA AGC TG) and vehicle served as controls. Oligonucleotides showed no relevant complementarity to other sequences after a computer database search. The oligonucleotides were purified by HPLC and Sephadex gel chromatography or with ion exchange chromatography (Qiagen™). Sterile pyrogen-free Ringer solution or aCSF were used as vehicle.

Bolus infusion of TH antisense oligonucleotide into the lateral ventricle

In a first set of experiments we performed bolus injections of complete phophorothioate TH antisense oligonucleotides into the lateral ventricle. By the i.c.v. route of administration we observed changes reminiscent of sickness responses (curled body posture, piloerection, immobilization, reduced food and water intake) in most of the experimental animals, in the phosphorothioate antisense, sense and mixed bases controls. These disturbing effects on the behavior of the animals were dose-dependent and hampered experimental designs. We were also able to demonstrate that i.c.v. administered complete phosphorothioate TH antisense and their respective sense and scrambled oligonucleotide controls produced motor disturbances in rats, which were associated with hypothermia (Schöbitz et al., 1997) directly after the administration. I.c.v. administration of endcapped phosphorothioate TH oligonucleotides produced the same reversible symptoms. The observed side-effects did not depend on the vehicle (saline, water, carbogen buffer, aCSF, etc.) or sequence used, the mode of modification or contamination of the probes by endotoxins, but seems to depend on the direct application of DNA into the brain (see 9.4.). However, the non-specific side-effects observed did not necessarily override specific effects, when compared to controls. Thus, despite a non-specific reduction of feeding behavior, we were able to demonstrate specific suppression of food-reinforced behavior by a TH antisense oligonucleotide caused by the inhibition of the central catecholaminergic system (Spanagel et al., 1998).

Bolus infusion of TH antisense into the ventral tegmental area

In another set of experiments we performed microinfusions of complete phosphoro-thioate TH antisense oligonucleotides into the ventral tegmental area. DAergic A10 neurons, which are known to be the essential part of the brain reinforcement system, are located in this brain region. We selected antisense oligonucleotides corresponding to the start coding region of rat TH mRNA and a mixed base oligonucleotide served as control. The oligonucleotides (5 μg/0.5 μl) were infused over a preimplanted guide cannula using a microinfusion pump into the ventral tegmental area. The solution was administered over a 30 sec period with the injection needle left in place for additional 30 sec to ensure complete delivery of the solution. The antisense effect was tested upon food-reinforced operant responding on a fixed ratio 10 (FR10) schedule. For this procedure rats were trained in standard operant chambers. After initial autoshap-ing the number of lever presses to obtain reinforcers of 45-mg food pellets was gradually increased to 10. This FR10 schedule was then maintained throughout the experiment. Training sessions in the Skinner boxes were conducted twice daily at 08:00 h and 20:00 h and lasted 20 min each. 36 h after TH antisense infusion operant behavior was markedly reduced, and this suppression was fully reversed within 5 days following the infusion (Skutella et al., 1994c). Accordingly, TH immunoreac-tivity in the VTA was reduced in comparison to control experiments using mixed base oligonucleotides (Skutella et al., 1994c). These results indicate that a single in-tramesencephalic TH antisense oligonucleotide injection was able to decrease the production of the functional enzyme resulting in suppressed operant behavior.

Chronic infusion (with osmotic minipumps) of TH antisense oligonucleotide into the substantia nigra

In a following study we used phosphorothioate and unmodified antisense oligonu-cleotide complementary to TH mRNA to examine whether its administration into the substantia nigra of the rat *in vivo* interferes significantly with physiological and be-havioral parameters as would be expected on the basis of known effects of conven-tional neurotoxic interventions. oligonucleotides were infused (0.5 μg/μl/h) unilater-ally into the substantia nigra by an osmotic minipump-system over 14 days. TH an-tisense oligonucleotide treatment was compared to previous findings with another agent known to deplete DA in this system, that is, the unilateral administration of 6-OHDA into the substantia nigra or ascending DA fibers. This lesion is an established experimental approach that has been widely employed to study DA function in the brain, to investigate the effects of drugs on central DA neurons, and to model critical elements of Parkinson's disease (for review see Reading and Dunnett, 1994). The ex-perimental lesion is known to destroy nigrostriatal DA neurons on one side of the brain. This is paralleled by specific behavioral deficits attributed to the hemispheric imbalance of DAergic activity within the basal ganglia characterized by turning be-havior towards the hemisphere with lower DA content and a bias of reactivity to stimuli presented to the contralateral side of the body (Ungerstedt, 1971; Steiner et al., 1989; Fornaguera et al., 1993). In addition to assess behavioral effects of intra-substantia nigra antisense oligonucleotide against TH, we characterized its effects on DA by *in vivo* microdialysis, examined its action on TH and glutamate decarboxylase mRNA by PCR and in situ hybridization and on TH and glutamate decarboxylase content by immunoassay.

The administration of complete phosphorothioate antisense oligonucleotide tar-geted against TH unilatarally into the substantia nigra produced unspecific turning behavior irrespective of the oligonucleotide sequence. When challenged with the in-direct DA agonist amphetamine (5mg/kg; i.p.), a contraversive asymmetry in turning towards the side of antisense infusion became evident on day 1 after minipump im-plantation. Contraversive turning was also observed without amphetamine challenge. This contraversive turning profile lasted until day 4 after the commencement of an-tisense infusion (figure 3A). Rats infused unilaterally with phosphorothioate sense or mismatch oligonucleotide or vehicle over 4 days showed the same asymmetry in turning behavior with respect to the side of infusion. This unspecific sequence-independent effect was dose-dependent. Thus, oligonucleotide doses lower than 0.1 µg/µl/h were without effect whereas a dose of 1 µg/µl/h even further increased unspe-cific contraversive turning behavior. In contrast, animals treated with unmodified oli-gonucleotide antisense showed ipsilateral turning behavior when challenged systemi-cally with amphetamine whereas mismatch- or vehicle-infused rats showed no such behavioral asymmetries. Indications for such an ipsiversive asymmetry became ob-servable on days 2 and 4 after the start of antisense-infusion, and were significant on day 6 and thereafter (until day 14) (figure 3B) (Skutella et al., 1997). Ipsilateral asymmetries in turning after amphetamine are characteristic effects of unilateral 6-OHDA lesions of the nigrostriatal DA system (Ungerstedt, 1971). Amphetamine, which requires endogenous DA for its action, is known to increase DA activity only in the intact (contralateral) hemisphere, thus exaggerating the DAergic imbalance between the denervated and intact side. This is behaviorally expressed as ipsiversive turning.

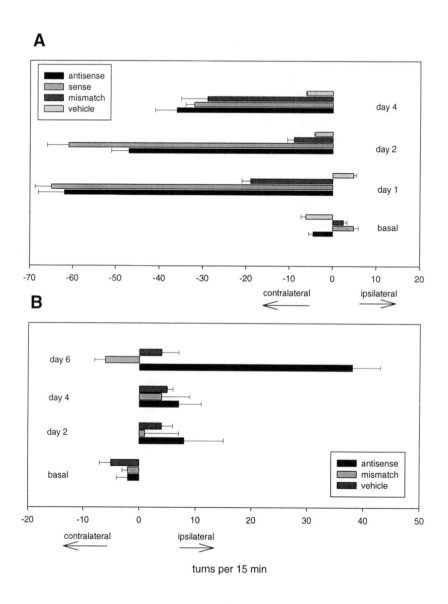

Figure 3. A Turning behavior after injections of 5mg/kg amphetamine (i.p.) on the day before (basal) and 1, 2 and 4 days after oligonucleotide or vehicle treatment. Animals were infused with complete phosphorothioate oligonucleotides (0.5 µg/µl/h) unilaterally into the substantia nigra. **B** Turning behavior after injections of 5mg/kg amphetamine (i.p.) on the day before (basal) and 2, 4 and 6 days after oligonucleotide or vehicle treatment. Animals were infused with unmodified oligonucleotides (0.5 µg/µl/h) unilaterally into the substantia nigra. Circling behavior is expressed as the difference of ipsilateral and contralateral turns (means + S.E.M).

In the substantia nigra, antisense treatment had no effects on TH mRNA. RNA analysis, via reverse transcriptase polymerase chain reaction followed by Southern Blot analysis, showed that TH mRNA levels in the substantia nigra were not markedly different between antisense- and mismatch-treated animals (Skutella et al., 1997). Possible RNA content fluctuations of the RNA preparations can be excluded, since the ß-actin RT-PCR consistently yielded equal amounts of PCR product under identical reaction conditions. *In situ* hybridization of TH and glutamate decarboxylase mRNA, which served as an internal control for the antisense effect, yielded no indication for reductions in radioactive signaling around the site of infusion in the substantia nigra in all experimental groups (Skutella et al., 1997). The unaffected mRNA levels are in contrast to observations with 6-OHDA lesions (Berod et al., 1987; Pasinetti et al., 1989; Sherman and Moody, 1995). However, it should be remembered that the 6-OHDA lesion destroys DAergic neurons, and thus their genetic material, whereas the antisense approach should not destroy the neuron and its genetic code, but should affect its expression. In this context it is interesting to note that other experiments using the *in vivo* antisense approach observed either an unchanged (Moat-Staats et al., 1993) or an elevated mRNA content (Landgraf et al., 1995) despite pronounced behavioral effects. Taken together, these results on the mRNA level favor the view that antisense oligonucleotide function via hybrid arrested translation, presumably in a large number of neurobiological applications. Clear molecular evidence for the proposed antisense mechanisms are rare and evidence for RNase-H action in the brain seems controversial. *In vitro* and *in vivo* experiments from our group and others showed that antisense oligonucleotide treatment in neuronal systems led to specific physiological and behavioral effects but, at the mRNA level no reduction of the corresponding message was observable, in fact the mRNA content was either clearly elevated or unchanged (Probst and Skutella , 1996). These results favor the view that antisense oligonucleotides might function via hybrid arrested translation rather than RNase-H driven mRNA cleavage. This means that during the translation from mRNA to protein, the application of antisense oligonucleotides establishes DNA-mRNA hybrids which interfere with the initiation or progress of the ribosomal apparatus causing an translational arrest without degrading the target mRNA. This mode of antisense effect is dependent on saturation of target mRNA with oligonucleotide and requires high molar concentration of antisense oligonucleotide to achieve the desired downregulation of the protein under observation.

TH and glutamate decarboxylase enzyme content in the unmodified antisense and mismatch oligonucleotide-infused substantia nigra as well as in the striatum were evaluated by immunoassay. Solid phase immunoassay for tyrosine hydroxylase revealed a significant reduction of immunoreactive peptide in homogenates of substantia nigra and striatum of antisense-treated rats at the infused brain site (ipsilateral) versus the non-infused brain site (contralateral). The mismatch-infused control group showed no differences in tyrosine hydroxylase protein content between the ipsilateral and contralateral sides. Glutamate decarboxylase immunoreactivity was unchanged in all experimental groups (Skutella et al., 1997).

Tissue levels of DA, measured in postmortem tissue punches of neostriatum and substantia nigra, were reduced in the oligonucleotide-treated hemisphere. Furthermore, basal extracellular levels of DA, monitored by in vivo microdialysis, were also lower in the neostriatum ipsilateral to antisense infusion and showed a weaker response to an amphetamine challenge when compared than the contralateral side. These effects were not observed after mismatch oligonucleotide or vehicle infusion into the substantia nigra. Finally, the GABAergic enzyme glutamate decarboxylase was not affected in the antisense-treated substantia nigra, indicating that unspecific damage in this area was not caused by this treatment.

In this experiment, the infusion of antisense oligonucleotide against TH mRNA in the substantia nigra led to a number of neurochemical- and subsequent behavioral changes related to the nigrostriatal DA system. Thus, turning behavior observed ipsilateral to the antisense infusion site was associated with reduced DA content and lower basal extracellular DA release in the nigrostriatal system. This neurochemical deficit of nigrostriatal DA was most likely related to decreased TH protein levels occuring in this system following antisense treatment. However, these effects were paralled by unchanged mRNA levels for TH, glutamate decarboxylase and ß-actin, which indicates ongoing transcriptional activity and strongly argues against the possibility of unmodified antisense oligonucleotide infusion producing non-specific effects in terms of neuronal toxicity and/or the general disruption of protein synthesis.

In general, the results of this study indicate the fearability of using unmodified antisense oligonucleotide to manipulate neurochemical systems in such a way as to allow their functional analysis. The effects on the mesolimbic and nigrostriatal system matched the known action of 6-OHDA toxicity in this system to the expected degree. It is likely that the technique will also be useful in targeting other enzymes that have escaped functional analysis due to the lack of selective pharmacological tools.

9.7 IN VIVO ANTISENSE TARGETING OF DA-RECEPTORS

Application of homology cloning techniques led to the identification of at least five different subtypes of DA-receptors (Civelli et al., 1993; Sokoloff and Schwartz, 1995). A novel approach to discriminate among closely related receptor subtypes such as the DA-receptor subtypes is the treatment with an antisense oligonucleotide which targets a specific receptor subtype mRNA. Recent studies have demonstrated the efficacy of antisense treatment in downregulating DA-receptor protein synthesis and interfering with function (for review see Weiss et al., 1997).

Due to its distribution in the brain, in particular in the DA midbrain system (Levesque et al., 1992; Murray et al., 1994) the D3-DA-receptor may be an appropriate target in the treatment of psychiatric disorders such as schizophrenia and drug addiction (for review see Shafer and Levant, 1998). Recently it could be shown that cocaine self-administration is modulated through D3-DA-receptors (Caine and Koob, 1993). In another study, long-term voluntary ethanol consumption led to specific changes in the D3-DA-receptor gene expression, whereas ethanol drinking did not alter the mRNA expression of D1-, D2-, D4- or D5-DA-receptors (Eravci et al., 1997). Further elucidation of the functional role of the D3-DA-receptor in the mediation of reinforcing effects induced by natural reinforcers such as food as well as drugs of abuse is warranted, but the appropriate studies are still severely hampered by the lack of selective D3-DA receptor antagonists.

It is also well known that neuroleptic drugs acting via the blockade of central DA-receptors disrupt the learning and performance of operant behavior motivated by a variety of positive reinforcers such as food, water, brain stimulation and drugs of abuse (Wise, 1982). In the following experiments we used the antisense approach to study the involvement of D3-DA-receptors in the mediation of food-reinforcement. Various D3-DA-receptor oligonucleotides were bilaterally infused via mini-osmotic pumps into the nucleus accumbens shell region, an area which has the highest density for these receptors and which is also closely linked to reinforcement processes (Pontieri et al., 1995). The sequence of the double endcapped, gel-filtered and HPLC purified phoshorothioate antisense oligonucleotide used in the present study were designed according to the primary structure of the mRNA which encodes for the D3-DA receptor. The effective antisense sequence (GAG GTG CCA TGG CCC ACA CAG) corresponded to the NH_2-terminus around the initiating codon. Sense sequence oligonucleotide (CTG TGT GGG CCA TGG CAC CTC), mismatch oligonucleotide as well as vehicle served as controls. oligonucleotides showed no relevant complementarity to other sequences after computer database search. For continuous infusion of oligonucleotides or vehicle control (sterilized Ringer; pH 7.4) into the nucleus accumbens shell region, each rat was implanted with a mini-pump device (figure 2), which was assembled prior to surgery and consisted of three parts: (1) an osmotic mini-pump (nominal infusion rate: 0.5 μl/h, Alzet model 2002), (2) a polyethylene tube (10 cm length; PE 800-100-200) formed into a loop and (3) an angled 27-gauge stainless steel cannula. Mini-pump, cannula and tube were filled with 240 μl of one of the oligonucleotide-containing solutions (0.5 μg/μl) or vehicle. Once filled, the tubing-cannula assembly was joined to the pump. All steps were performed under sterile conditions. Using stereotactic surgery, the cannulae were lowered into the left and right nucleus accumbens shell region of pento-barbiturate-anesthetized rats. The cannulae were secured to the skull with two stainless steel screws and dental cement. The osmotic mini-pumps were placed between the scapulae in a small subcutaneous cavity. The incision was closed with metal clips and the rats were then allowed to recover for several days before testing. To prevent infections, all animals were subcutaneously injected with an antibiotic. Subsequently, the antisense effect was tested upon food-reinforced operant responding on a FR10 schedule. For this purpose rats were trained in standard operant chambers prior surgery. After initial autoshaping the number of lever presses to obtain reinforcers of 45-mg food pellets was gradually increased to 10. This FR10 schedule was then maintained throughout the experiment. As soon as the animals reached the criterion for stable responding in between sessions, rats were implanted with the minipump device as described and delivery of the antisense oligonucleotide or the respective control solutions began immediately.

Five days after D3-DA-receptor antisense oligonucleotide infusion operant behavior was markedly reduced (figure 4). Accordingly, [^{3}H]-(+)-7-OH-DPAT-binding
was significantly reduced in the nucleus accumbens shell, but not in other regions. In
comparison, vehicle and the appropriate oligonucleotide controls had no influence on
D3-DA receptor density (figure 5). The selectivity of the antisense oligonucleotide
treatment was further confirmed by unchanged D1- and D2-DA-receptor densities. In
a similar approach, a reduction of [^{3}H]-spiperone binding in the limbic forebrain and
an accompanied increase in DA synthesis by antisense oligonucleotide treatment was
found (Nissbrandt et al., 1995). These results show that the chronic bilateral infusion
of endcapped phosphorothioate D3-DA-receptor antisense oligonucleotide into the
nucleus accumbens shell region leads to a selective reduction of this DA-receptor
subtype which is accompanied by suppressed food-reinforced behavior. This finding
further highlights the role of D3-DA-receptors in reinforcement processes.

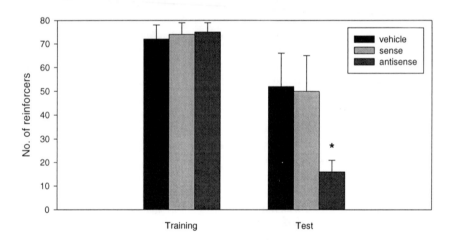

Figure 4. The effects of D3-DA-receptor antisense, sense and vehicle infusions into the nucleus
accumbens shell region upon food-reinforced operant responding on a FR 10 schedule. Ordinate gives the
numbers of reinforcers (food pellets). Each bar represents the number of reinforcements of 6 animals +
S.E.. Asterisks indicate significant differences to baseline responding as well as to sense and vehicle
treatment; P<0.01.

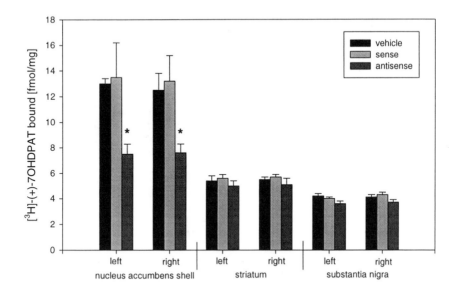

Figure 5. Specific binding of [^{3}H]-(+)-7-OHDPAT (fmol/mg protein) in several brain structures. Binding was performed in rat brains which have received chronic intra-nucleus accumbens shell infusions of vehicle, sense and antisense oligonucleotides (n = 4-6 per group), respectively. Asterisks indicate significant differences to the vehicle and sense control; * P<0.01.

9.8 CONCLUSIONS

In vivo antisense targeting supplements our repertoire of pharmacological tools to study the involvement of a certain gene product in physiological and pathological processes although specificity still remains a problem. Our data presented here suggest at least to use oligonucleotide analogs with minor deviation from the native oligonucleotide structure. In particular complete phosphorothioate modifications introduce pharmacological consequences *in vivo* that do not depend on strict complementarity of their sequences to target mRNAs. This might result in several unspecific side-effects which can even overshadow specific behavioral alterations and thus can lead to misinterpretations. To minimize side-effects it is reasonable to infuse i.c.v endcapped oligonucleotides, since phosphorodiesters are rapidly degraded in the intraventricular compartment. Although these and other guidelines will help to optimize antisense efficacy while maintaining maximum selectivity, the experimental conditions and the design have to be reevaluated for each new study.

9.9 REFERENCES

Aggrawal BB, Schwarz L, Hogan ME, Rando RF. Triple helix-forming oligodeoxyribonucleotides targeted to the human tumor necrosis factor (TNF) gene inhibit TNF production and block the TNF-dependent growth of human glioblastoma tumor cells. *Cancer Res* 1996;56:5156-64

Arar K, Aubertin AM, Roche AC, Monsigny M, Mayer R. Synthesis and antiviral activity of peptide-oligonucleotide conjugates prepared by using N alpha-(bromoacetyl)peptides. *Bioconjugate Chem* 1995;6:573-77

Baker BF, Lot SS, Condon TP, Cheng-Flournoy S, Lesnik EA, Sasmor HM, Bennett CF. 2'-O-(2-Methoxy)-ethyl-modified anti-intercellular adhesion molecule 1 (ICAM-1) oligonucleotides selectively increase the ICAM-1 mRNA level and inhibit formation of the ICAM-1 translation initiation complex in human umbilical vein endothelial cells. *J Biol Chem* 1997;272:11994-2000

Basu S, Wickstrom E. Synthesis and characterization of a peptide nucleic acid conjugated to a D-peptide analog of insulin-like growth factor 1 for increased cellular uptake. *Biocon Chem* 1997;8:481-88

Beck GF, Irwin WJ, Nicklin PL, Akhtar S. Interactions of phosphodiester and phosphorothioate oligonucleotides with intestinal epithelial caco-2 cells. *Pharma Res* 1996;13:1028-37

Berod A, Faucon Biguet N, Dumas S, Bloch B, Mallet J. Modulation of tyrosine hydroxylase gene expression in the central nervous system visualized by in situ hybridization. *Proc Natl Acad Sci USA* 1987;84:1699-703

Caine SB, Koob GF. Modulation of cocaine self-administration in the rat through D-3 dopamine receptors. *Science* 1993;260:1814-16

Cirino NM, Li G, Xiao W, Torrence PF, Silverman RH. Targeting RNA decay with 2',5' oligoadenylate-anti-sense in respiratory syncytial virus-infected cells. *Proc Natl Acad Sci USA* 1997;94:1937-42

Civelli O, Bunzow JR, Grandy DK. Molecular diversity of dopamine receptors. *Annu Rev Pharmacol Toxicol* 1993; 32:281-307

Eravci M, Großpietsch T, Pinna G, Schulz O, Kley S, Bachmann M, Wolffgramm J, Götz E, Heyne A, Meinhold H, Baumgartner A. Dopamine receptor gene expression in an animal model of behavioral dependence on ethanol. *Mol Brain Res* 1997;50:221-29

Fornaguera J, Schwarting RKW, Boix F, Huston JP. Behavioral indices of moderate nigrostriatal 6-hydroxydopamine lesion: a preclinical Parkinson's model. *Synapse* 1993;13:179-85

Hebb MO, Robertson HA. End-capped antisense oligodeoxynucleotides effectively inhibit gene expression in vivo and offer a low-toxicity alternative to fully modified phosphorothioate oligodeoxynucleotides. *Mol Brain Res* 1997;47:223-28

Heidenreich O, Gryaznov S, Nerenberg M. RNase H-independent antisense activity of oligonucleotide N3 '--> P5 ' phosphoramidates. *Nucl Acids Res* 1997;25:776-80

Ho SP, Britton DH, Stone BA, Behrens DL, Leffet LM, Hobbs FW, Miller JA, Trainor GL. Potent antisense oligonucleotides to the human multidrug resistance-1 mRNA are rationally selected by mapping RNA-accessible sites with oligonucleotide libraries. *Nucl Acids Res* 1996;24:1901-07

Landgraf R. Antisense targeting in behavioural neuroendocrinology. *J Endocrinol* 1996;151:333-40

Landgraf R, Naruo T, Vecsernyes M, Neumann I. Neuroendocrine and behavioral effects of antisense oligo-nucleotides. *Eur J Endocrinol* 1997;137:326-35

Landgraf R, Gerstberger R, Montkowski A, Probst JC, Wotjak CT, Holsboer F, Engelmann M. V1 Vasopres-sin receptor antisense oligodeoxynucleotide into septum reduces vasopressin binding, social discrimination abilities, and anxiety-related behavior in rats. *J Neurosci* 1995;15:4250-58

Lévesque D, Diaz J, Pilon C, Martres M-P, Giros B, Souil E, Schott D, Morgat J-L, Schwartz J-C, Sokoloff P. Identification, characterization, and localization of the dopamine D_3 receptor in rat brain us-ing 7-[^{3}H]hydroxy-*N,N*-di-*n*-propyl-2aminotetralin. *Proc Natl Acad Sci USA* 1992;89:8155-59

Li B, Hughes JA, Phillips MI. Uptake and efflux of intact antisense phosphorothioate deoxyoligonucleo-tide directed against angiotensin receptors in bovine adrenal cells. *Neurochem Int* 1997;31:393-403

Loke SL, Stein CA, Zhang XH, Mori K, Nakanishi M, Subasinghe C, Cohen JS, Neckers LM. Characteri-zation of oligonucleotide transport into living cells. *Proc Natl Acad Sci USA* 1989;86:3474-78

Mergny JL, Duval-Valentin G, Nguyen CH, Perrouault L, Faucon B, Rougee M, Montenay-Garestier T, Bisagni E, Helene C. Triple helix-specific ligands. *Science* 1992;256:1681-84

Moat-Staats BM, Retsch-Bogart GZ, Price WA, Jarvis HW, D'Ercole AJ, Stiles AD. Insulin-like growth fac-tor-I (IGF-I) antisense oligodesoxynucleotide mediated inhibition of DNA synthesis by Wi-38 cells: evidence for autocrine actions of IGF-I. *Mol Endocrinol* 1993;7:171-80

Monia BP, Sasmor H, Johnston JF, Freier SM, Lesnk EA, Muller M, Altmann K-H, Moser H, Fabbro D. Sequence specific antitumor activity of a phosphorothioate oligodeoxyribonucleotide targeted to human c-raf kinase supports an antisense mechanism of action in vivo. *Proc Natl Acad Sci USA* 1996b;93: 15481-85

Monia BP. First- and second-generation antisense inhibitors targeted to human c-raf kinase: in vitro and in vivo studies. *Anti-Cancer Drug Des* 1997;12:327-41

Moulds C, Lewis JG, Froehler BC, Grant D, Huang T, Milligan JF, Matteucci MD, Wagner RW. Site and mechanism of antisense inhibition by C-5 propyne oligonucleotides. *Biochemistry* 1995;34:5044-53

Murray AM, Ryoo HL, Gurevich E, Joyce JN. Localization of dopamine D_3 receptors to mesolimbic and D_2 receptors to mesostriatal regions of human forebrain. *Proc Natl Acad Sci USA* 1994;91:11271-75

Neumann I, Kremarik P, Pittman QJ. Acute, sequence-specific effects of oxytocin and vasopressin an-tisense oligonucleotides on neuronal responses. *Neuroscience* 1995;69:997-1003

Nielsen PE, Egholm M, Berg RH, Buchardt O. Sequence-selective recognition of DNA by strand displace-ment with a thymine-substituted polyamide. *Science* 1991;254:1497-500

Nissbrandt H, Ekman A, Eriksson E, Heilig M. Dopamine D3 receptor antisense influneces dopamine synthe-sis in rat brain. *Neuroreport* 1995;6:573-76

Pasinetti GM, Lerner SP, Johnson SA, Morgan DG, Telford NA, Finch CE. Chronic lesions differentially de-crease tyrosine hydroxylase messenger RNA in dopaminergic neurons of the substantia nigra. *Mol Brain Res* 1989;5:203-09

Peyman A, Uhlmann E. Minimally modified oligonucleotides - combination of end-capping and pyrimidine-protection. *Biol Chem* 1996;377:67-70

Pontieri FE, Tanda G, Di Chiara G. Intravenous cocaine, morphine, and amphetamine preferentially increase extracellular dopamine in the shell as compared with the core of the rat nucleus accumbens. *Proc Natl Acad Sci USA* 1995;92:12304-08

Probst JC, Skutella T. Elevated messenger RNA levels after antisense oligodeoxynucleotide treatment in vitro and in vivo. *Biochem Biophys Res Com* 1996;225:861-68

Probst JC, Skutella T. The G-tetrad in antisense targeting. *Trends Genet* 1996;12: 290-91

Reading PJ, Dunnett SB. "6-Hydroxydopamine lesions of nigrostriatal neurons as an animal model of Parkinson´s disease". *Toxin-induced Models of Neurological Disorders.* 1994. Woodruff ML, Nonnemann AJ, eds. Plenum Press, New York

Reed MW, Fraga D, Schwartz DE, Scholler J, Hinrichsen RD. Synthesis and evaluation of nuclear targeting peptide-antisense oligodeoxynucleotide conjugates. *Bioconjugate Chem* 1995;6:101-08

Schmidt A, Sindermann J, Peyman A, Uhlmann E, Will DW, Muller JG, Breithardt G, Buddecke E. Sequence-specific antiproliferative effects of antisense and end-capping-modified antisense oligodeoxynucleotides targeted against the 5'-terminus of basic-fibroblast-growth-factor mRNA in coronary smooth muscle cells. *Eur J Biochem* 1997;248:543-49

Schöbitz B, Pezeshki G, Probst JC, Reul JM, Skutella T, Stöhr T, Holsboer F, Spanagel R. Centrally administered oligodeoxynucleotides in rats: occurrence of non-specific effects. *Eur J Pharm* 1997;331:97-107

Shafer RA, Levant B. The D₁ dopamine receptor in cellular and organismal function. *Psychopharmacology* 1998;135:1-16

Sherman TG, Moody CA. Alterations in tyrosine hydroxylase expression following partial lesions of the nigrostriatal bundle. *Mol Brain Res* 1995;29:285-96

Skorski T, Perrotti D, Nieborowska-Skorska M, Gryaznov S, Calabretta B. Antileukemia effect of c-myc N3'-->P5' phosphoramidate antisense oligonucleotides in vivo. *Proc Natl Acad Sci USA* 1997;94:3966-71

Skutella T, Probst JC, Engelmann M, Wotjak CT, Landgraf R, Jirikowski GF. Vasopressin antisense oligonucleotide induces temporary diabetes insipidus in rats. *J Neuroendocrinol* 1994a;6:626-31

Skutella T, Stöhr T, Probst JC, Ramalho-Ortigao FJ, Holsboer F, Jirikowski GF. Antisense oligodeoxynucleotides for in vivo targeting of corticotropin-releasing hormone mRNA: comparison of phosphorothioate and 3'-inverted probe performance. *Hormone Metabol Res* 1994b;26:460-64

Skutella T, Probst JC, Jirikowski GF, Holsboer F, Spanagel R. Ventral tegmental (VTA) injections of tyrosine hydroxylase phosphorothioate antisense oligonucleotide suppress schedule-controlled behavior in rats. *Neurosci Lett* 1994c;167:55-58

Skutella T, Schwarting RKW, Huston JP, Sillaber I, Probst JC, Holsboer F, Spanagel R. Infusions of tyrosine hydroxylase antisense oligodesoxynucleotides into the substantia nigra in vivo: effects on THmRNA, TH protein content, striatal dopamine release and behavior. *Eur J Neurosci* 1997;9:210-21

Skutella T, Probst JC, Renner U, Holsboer F, Behl C. Corticotropin- releasing hormone receptor (type I) antisense targeting reduces anxiety. *Neuroscience* 1998;in press

Sokoloff P, Schwartz J-C. Novel dopamine receptors half a decade later. *Trends Pharmacol Sci* 1995;16:270-75

Spanagel R, Schöbitz B, Engelmann M. "Non-specific-effects of centrally administered oligonucleotides". *Modulating Gene Expression by Antisense oligonucleotides to Understand Neural Function*. 1998. Mc Carthy MM, ed. Kluwer Academic Publishers, Norwell, MA

Standifer KM, Chien CC, Wahlestedt C, Brown GP, Pasternak GW. Selective loss of delta opioid analgesia and binding by antisense oligodeoxynucleotides to a delta opioid receptor. *Neuron* 1994;12:805-10

Stein CA, Krieg AM. Problems in interpretation of data derived from *in vitro* and *in vivo* use of antisense oligodeoxynucleotides. *Antisense Res Dev* 1994;4:67-69.

Steiner H, Bonatz AE, Huston JP, Schwarting RKW. Lateralized wall-facing versus turning as measures of behavioral asymmetries and recovery of function after injection of 6-hydroxydopamine into the substantia nigra. *Exp Neurol* 1989;99:556-66

Torrence PF, Maitra RK, Lesiak K, Khamnei S, Zhou A, Silverman RH. Targeting RNA for degradation with a (2'-5')oligoadenylate-antisense chimera. *Proc Natl Acad Sci USA* 1993;90:1300-04

Ungerstedt U. Striatal dopamine release after amphetamine or nerve degeneration revealed by rotational behaviour. *Acta Physiol Scand* 1971;82:49-68

Weiss B, Zhang SP, Zhou LW. Antisense strategies in dopamine receptor pharmacology. *Life Sci* 1997;60:433-55

Wise RA. Neuroleptics and operant behavior: the anhedonia hypothesis. *Behav Brain Sci* 1982;5:39-87

10 Lipid-Based Carriers for the Systemic Delivery of Antisense Drugs

Murray S. Webb, Sandra K. Klimuk,
Sean C. Semple and Michael J. Hope

Inex Pharmaceuticals Corporation
Burnaby, British Columbia,
Canada

10.1 INTRODUCTION

Why use carriers for antisense oligonucleotides

Antisense oligonucleotides are now widely recognized as a new class of therapeutic agents with exceptional potential for achieving biological efficacy with high target specificity. Intense effort in recent years has resulted in a number of antisense oligonucleotides currently in human clinical trials and one recently approved by the FDA for the treatment of CMV retinitis. All of these trials involve the parenteral administration of free (unencapsulated) oligonucleotide, and there is mounting evidence at the preclinical and clinical levels that antisense drugs are efficacious against a variety of diseases. Consequently, it might be asked "What is the rationale for the use of carriers to deliver antisense oligonucleotides *in vivo*?".

It is anticipated that delivery in lipid-based carriers represents an effective means for increasing the biological activity, and simultaneously decreasing non-specific toxicities, of antisense drugs. Specifically, given the experience gained through delivery of conventional drugs in liposomes, it is expected that lipid-based delivery of oligonucleotides will be effective at:

- Increasing oligonucleotide circulation lifetimes.
- Enabling bolus administration and/or decreased administration frequency.
- Increasing accumulation of oligonucleotide at sites of disease.
- Increasing drug potency.

- Increasing intracellular delivery in some tissues.
- Reducing class toxicities for phosphorothioate oligonucleotides.
- Protecting susceptible antisense chemistries from nuclease degradation.

In practice, all of the above combine to increase the therapeutic indices of antisense drugs in lipid-based delivery systems. In this chapter we will outline the benefits conferred to conventional pharmaceuticals by liposomal encapsulation to illustrate the key characteristics of liposomal delivery systems. With this as background, we provide a critical literature review of lipid-based formulations employed for the *in vivo* delivery of antisense oligonucleotides (Section 10.2). This is followed (Section 10.3) by a summary of the results we have observed *in vivo* for liposomal antisense directed against ICAM-1. In Section 10.4 we provide a description of a rationally-designed antisense carrier, the Stabilized Antisense-Lipid Particle (SALP), for systemic administration of oligonucleotides directed against cancer and inflammatory diseases. It will be demonstrated that the therapeutic activities of antisense oligonucleotides delivered by liposomal carriers can be enhanced significantly over the unencapsulated antisense drug. Furthermore, the biological benefits that are achieved by liposomal delivery are substantially influenced by details of the carrier design.

Background: Lipid carriers for the delivery of conventional drugs

Liposomal delivery systems have been widely employed to alter the pharmacokinetics, biodistribution, disease site accumulation and biological activities of diverse therapeutic drugs. For many conventional chemotherapeutic agents, the influence of liposomal encapsulation is profound, increasing drug circulation times from several minutes to many hours. Associated with this increase in circulation lifetime is an increase in the accumulation of both lipid and drug in organs of the reticuloendothelial systems (RES), primarily the liver and spleen. Liposomal drug carriers also have the ability to accumulate passively at sites where the vascular barrier has been compromised by disease progression. Examples include ascitic (Boman *et al.* 1994; Mayer *et al.*, 1990c; Webb *et al.*, 1995) and a variety of solid tumors (Mayer *et al.*, 1990b; Parr *et al.*, 1997; Webb *et al.*, 1995), pulmonary bacterial infections (Bakker-Woudenberg *et al.*, 1992), and sites of inflammation (Boerman *et al.* 1997).

Liposomal vincristine, ciprofloxacin and doxorubicin will be used to illustrate the pharmacodynamic benefits associated with encapsulation of drugs in lipid-based carriers. For more detailed reviews, the reader is referred to Boman *et al.* (1997) and Chonn and Cullis (1995). Vincristine is the drug of choice in managing childhood leukemia and is also a key component in the treatment regime for Hodgkin's disease and non-Hodgkin's lymphomas. Doxorubicin (adriamycin) is one of the most commonly used antineoplastic agents (Tardi *et al.*, 1996) with activity against a number of solid tumors. Ciprofloxacin is a synthetic fluoroquinolone antibiotic with broad spectrum efficacy against a wide variety of bacteria, including *Staphylococcus aureus, Pseudomonas aeruginosa, Klebsiella pneumoniae, Mycobacterium tuberculosis* and *Mycobacterium avium* complex (Fenlon and Cynamon, 1986; Gay *et al.*, 1984; Sanders *et al.*, 1987).

Encapsulation of vincristine in liposomal delivery systems increases the circulation lifetime of the drug by at least 50-fold, with the t 1/2 values increasing from 0.16 h to $\geq$ 8.0 h for various liposomal formulations (Webb *et al.*, 1995). This represents $\geq$ 30-fold increases in plasma drug concentrations as a result of liposomal encapsulation. Similar results have been shown for doxorubicin (Bally *et al.*, 1990) and ciprofloxacin (Webb *et al.*, 1998a). For compounds that readily diffuse through liposomal membranes, such as vincristine and ciprofloxacin, the lipid composition can be a crucial regulator of drug retention in the circulation (Webb et al, 1998a; Webb *et al.*, 1995). It is important to note that increases in dose intensity at a disease site are achieved when the liposomal carrier, containing the encapsulated drug, extravasates from the circulation to tumor interstitium (Webb *et al.*, 1995). This mechanism is distinct from prolonged exposure of tumors to drug that is slowly leaked from liposomes that remain in the circulation. Since the liposomal carrier, and its encapsulated drug, extravasate as a unit from the circulation to solid tumors, excessive leakage in the circulation will result in the accumulation of partially empty carriers at the disease site. It has been experimentally demonstrated that more permeable carriers lead to a reduction in the delivered drug dose and, consequently, sub-optimal therapy (Webb *et al.*, 1995). On the other hand, for drugs which are relatively impermeable, alterations in the lipid composition can have a relatively small impact on both pharmacokinetic properties and biological activity (Mayer *et al.*, 1990a). In cases where a membrane impermeable drug is encapsulated by a liposomal membrane with a very low permeability coefficient, the drug is not released from the carrier, does not become bioavailable and has negligible activity (Boman and Webb, unpublished). One might expect polyanionic antisense drugs to fall into this category as they do not diffuse through liposomal membranes (Vlassov *et al.*, 1994) and therefore carriers must be degraded in order to release their payload.

Additional design issues relevant to the delivery of antisense oligonucleotides include both the size and charge of the carrier/drug complex. Analysis of the effect of the vesicle diameter on the circulation lifetimes of neutral phosphatidylcholine/cholesterol (PC/chol) liposomes has shown that carriers with diameters greater than about 150 nm exhibit both reduced circulation lifetimes (Mayer *et al.*, 1989), reduced accumulation at extravascular tumor tissue (Bally *et al.*, 1994) and reduced antitumor activity (Mayer *et al.*, 1993; Mayer *et al.*, 1989). Moreover, many liposomal delivery systems bearing either net negative or net positive charge are rapidly removed by the RES after intravenous administration (Liu *et al.*, 1995; Zalipsky *et al.*, 1994) compared to net neutral delivery systems, a characteristic not ideal for the treatment of systemic disease. Formulation of plasmid or oligonucleotides with cationic liposomes is a strategy commonly used to enhance the delivery and nuclease stability of these genetic drugs (see Section 10.2. and Hope *et al.* 1998 for a review). However, this type of formulation leads to the worst of both worlds: large complexes with polydisperse size distribution and bearing a significant net charge. Not unexpectedly, these structures have poor circulation lifetimes after intravenous administration (Litzinger *et al.*, 1996; Osaka *et al.*, 1997; Zalipsky *et al.*, 1994) and may only be suitable for therapies employing local administration such as intratumoral injection, intraperitoneal and intratracheal administration.

Key characteristics of lipid-based delivery vehicles

The following is a list of key characteristics that an antisense delivery system will most likely have to exhibit for it to enhance the therapeutic index of antisense drugs:

High drug/lipid ratio. This ensures a maximal therapeutic benefit per delivery vehicle accumulating at a disease site. In addition, a high drug/lipid ratio will minimize the amount of lipid co-administered with the antisense drug.

High encapsulation efficiency. Will minimize the loss of oligonucleotide during manufacturing, making a product more cost effective.

Complete protection of drug from serum. In order to prevent the loss of active drug through degradation by extracellular nucleases, the oligonucleotide must either be completely encapsulated by a vesicle membrane or form a lipid-nucleic acid complex that achieves a high level of protection until the drug reaches its target tissue.

Mean diameter 100-150 nm with low polydispersity. Maximizes the circulation lifetime for distribution to systemic disease and facilitates accumulation at extravascular disease sites.

Minimal surface charge. A surface charge may be desirable for some applications; however, a net neutral surface maximizes the residence time of particles in the blood.

Drug retention optimized for *in vivo* application. The oligonucleotide should become bioavailable after the carrier accumulates at the disease site. In the ideal carrier, the oligonucleotide will be released into the cytoplasm from an endosomal compartment by a mechanism that can be regulated by alterations in carrier composition.

In the next section we summarize published studies concerning the *in vivo* delivery of antisense oligonucleotides employing lipid-based carriers.

10.2 STATE-OF-THE-ART: LIPID-BASED SYSTEMS FOR *IN VIVO* DELIVERY OF ANTISENSE OLIGONUCLOTIDES

In vitro experiments have clearly demonstrated that the vast majority of cells are essentially impermeable to free oligonucleotides. To achieve intracellular delivery of oligonucleotides, investigators have employed methods that transiently permeabilize the cell membrane. These methods include electroporation (Bergan *et al.*, 1993; Spiller *et al.*, 1998), direct microinjection (Buhr *et al.*, 1996; Clarenc *et al.*, 1993; Flanagan and Wagner, 1997; Moulds *et al.*, 1995) or streptolysin-O treatment (Giles *et al.*, 1998; Spiller et al, 1998). However, lipofection is the most widely used method to achieve intracellular delivery of antisense molecules *in vitro*. This technique employs cationic liposomes to form lipid:oligonucleotide complexes which are endocytosed and subsequently disrupt the endosome, releasing their contents into the cytoplasm (Crooke, 1995). As will be seen below, this mode of delivery is not suitable for *in vivo*, systemic applications.

Zwitterionic liposomal carriers

A summary of studies published in peer-reviewed journals describing the *in vivo* evaluation of antisense oligonucleotides in lipid-based delivery systems is presented in Table 1. Of the studies listed, liposomes with a net neutral (i.e. zwitterionic) surface charge were used in only two studies. One of these studies (Wielbo *et al.*, 1996), describes a formulation in which oligonucleotides were passively encapsulated in neutral PC/chol vesicles with final diameters in the range of 100-150 nm. The size and neutral surface charge of this carrier is likely to result in relatively long circulation lifetimes after administration. While the investigators did not specifically determine the circulation longevity or stability of the formulation *in vivo*, they reported enhanced accumulation of fluorescent-labeled oligonucleotide in the livers, compared to the unencapsulated oligonucleotide. Furthermore, liposomal antisense against angiotensinogen was more effective than encapsulated scrambled control or free antisense at decreasing both arterial blood pressures and plasma levels of angiotensin II. Passive encapsulation methodology gives rise to low encapsulation efficiency and low drug/lipid ratios (Semple *et al.*, manuscript in preparation). Consequently, Wielbo *et al.* (1996) were obliged to administer formulations at very low drug/lipid ratios of 0.002 (wt/wt). That is, to achieve an oligonucleotide dose of 0.2 mg/kg in rats, a lipid dose of 100 mg/kg was administered. While this lipid dose is typically well tolerated, achieving modest increases in the antisense doses would require increasing the lipid dose to very high levels. It should be noted that these investigators did not explicitly describe the phosphatidylcholine used in their studies. This detail may be important since it is known that acyl chain length and unsaturation can have a substantial impact on the circulation lifetime of liposomes (Allen *et al.*, 1989).

Tari *et al.* (1998) found that P-ethoxy oligonucleotides associate with neutral liposomes composed of dioleoylphosphatidylcholine (DOPC) with high efficiency (> 95%). Like methylphosphonates, P-ethoxy oligonucleotides do not carry a negative charge and are relatively lipid soluble. Presumably, the oligonucleotide is associated with the lipid bilayer rather than being in solution within the aqueous core of the vesicles. The formulation had no adverse toxicities at low antisense doses and exhibited circulation lifetimes greater than those of unencapsulated phosphorothioate or methylphosphonate oligonucleotides (Table 1). Comparisons were not made with unencapsulated P-ethoxy oligonucleotides due to its insolubility in aqueous buffer. The mean diameter of the liposomes employed was approximately 0.9 μm, consequently the circulation halflife for these particles would be considerably less than for a formulation with a mean diameter of about 0.1 μm. There are reports (Tari and Lopez-Berestein, 1997) that liposomal preparations of P-ethoxy oligonucleotides can be sized down by extrusion (Hope *et al.*, 1985), but these systems have yet to be evaluated *in vivo*.

Carriers for delivery of antisense drugs

TABLE 1: Summary of the published *in vivo* evaluations of antisense oligonucleotides in lipid-based delivery systems

Liposome Composition	Size (nm)	Gene target	Encap. Effic. (%)	Admin. Route; T1/2α (h)	Notes	Ref.
Cationic lipid:antisense complexes						
DOTAP:15mer PS oligonucleotide	Large[1]	TGFβ2	100[1]	Intratumoral; N/A	Delayed AC29 s.c. tumor growth; decreased TGF β2 mRNA	Marzo et al., 1997
DOSPER:13.5 Kb antisense plasmid	Large[1]	*Neu*	100[1]	Intratumoral; N/A	Decreased tumor growth; decreased *neu* mRNA & protein	Sacco et al., 1998
PC/chol/DDAB (3.2/1.6/1):15mer PO antisense[2]	1-10 μm	Raf-1	> 90	i.v.; 0.41	Decreased raf-1 protein levels, no tumor growth data	Gokhale et al., 1997
DOTMA/DOPE:20mer PS oligonucleotide (Lipofectin)	Large[1]	p120	100[1]	i.v.; 0.23	Free oligo = Lipofectin:oligo. i.v. Lipofectin increased uptake i.p.	Saijo et al., 1994
DOTMA/DOPE:20mer PS oligonucleotide (Lipofectin)	Large[1]	p120	100[1]	i.p.; N/A	Efficacy of Lipofectin:oligo >> free oligonucleotide	Perlaky et al., 1993
DC-chol/DOPE (1/1):18mer PS oligonucleotide	1.8 μm	None	100[1]	i.v.; Very short[3]	Rapid (5 min) accumulation in lungs; long term accumul. in liver	Litzinger et al., 1996
Net neutral liposomes						
DOPC; 18mer P-ethoxy oligonucleotide	≤ 900 nm	Various	> 95	i.v.; 0.11	Increased circulation halflife	Tari et al., 1998
PC/cholesterol (8/2): 18mer PS oligonucleotide	≈ 100 nm[1]	Angiotensinogen	< 10[1]	intra-arterial; ND	Decreased plasma AngII levels, decreased blood pressures	Wielbo et al., 1996

[1] not determined directly; estimated from preparation procedure.
[2] oligodeoxyribonucleotide; diester linkages with terminal phosphorothioate at 5' and 3' ends.
[3] not readily quantifiable; 80% of injected dose was present in the lungs at 5 min post-administration.

Anionic liposomal carriers

Very few investigators have attempted to encapsulate oligonucleotides in anionic liposomal carriers, probably because of the expected electrostatic repulsion between the antisense drug and the liposome surface. One system that has been evaluated extensively *in vivo* is a complex formed between Hemagglutinating Virus of Japan (HVJ) and anionic liposomes composed of PC, cholesterol and phosphatidylserine (Aoki *et al.*, 1997; Hangai *et al.*, 1998; Tomita *et al.*, 1998; Yamada *et al.*, 1996). This unusual formulation is prepared by passive encapsulation of oligonucleotides in the anionic liposomes, which are then sonicated and complexed to inactivated HVJ. Oligonucleotides encapsulated in the HVJ-liposome formulation have been administered by direct injection into the rat heart (Aoki *et al.*, 1997), intravitreally in mice (Hangai *et al.*, 1998), into the liver by either portal vein injection or direct injection (Tomita *et al.*, 1995), into the rat hypothalmus or lateral cerebroventricle by infusion (Yamada *et al.*, 1996) or intraluminal delivery into injured rat carotid arteries (Morishita *et al.*, 1994a,b). Formulation in the HVJ-liposome complex prolonged the duration of fluorescent oligonucleotides in the treated tissue and increased fluorescence detectable in cell nuclei. Tomita *et al.* (1995), reported sequence-specific decreases in blood pressure, plasma angiotensinogen concentrations and hepatic angiotensinogen mRNA *in vivo* using HVJ-liposomes. Similarly, Morishita *et al.* (1994a,b) treated ballon-injured rat carotid arteries with oligonucleotides against cdc2 kinase and cyclin B1 (Morishita *et al.*, 1994a) and cdk 2 kinase (Morishita *et al.*, 1994b) using HVJ-liposomes and observed sequence specific inhibition of neointimal formation.

Complexes between cationic lipids and antisense oligonucleotides

Most *in vivo* studies have employed oligonucleotides electrostatically complexed to cationic lipids or to mixtures of cationic lipids and various neutral lipids. Commonly employed cationic lipids include DOTMA (Perlaky *et al.*, 1993; Saijo *et al.*, 1994), DOTAP (Marzo *et al.*, 1997), DOSPER (Sacco *et al.*, 1998), DDAB (Gokhale *et al.*, 1997), DC-chol (Litzinger *et al.*, 1996) or DODAC (Hope *et al.*, 1998). Typically these lipids are combined with the fusogenic lipid dioleoylphosphatidylethanolamine (DOPE) (Hope *et al.*, 1998). In most protocols, liposomes containing cationic lipids are prepared in the absence of oligonucleotide by standard hydration and sonication methodologies to generate small unilamellar vesicles. Addition of polyanionic oligonucleotide triggers electrostatic aggregation to form a complex of liposomes and antisense that is both large (typically > 1-2 μm) and polydisperse (Table 1). The electrostatic association is responsible for "encapsulation efficiencies" approaching 100% (Table 1). However, the particles formed are highly unstable and the antisense drug is only partially protected (Hope *et al.*, 1998). It should be noted that oligonucleotides formulated with cationic lipids are typically electrostatically bound to the lipid surface and are not "encapsulated" per se. Rather, oligonucleotides formulated in anionic or zwitterionic liposomes will be present in the aqueous core of the vesicle and represent true encapsulated liposomal systems.

For *in vitro* delivery, complexes are typically made with a molar excess of cationic lipid over oligonucleotide phosphate groups so that the final particle has a net positive charge. This promotes association with the cell surface and endocytosis, particularly in the absence of serum proteins. But the combination of large size and residual charge leads to very rapid clearance of oligonucleotide and lipid from the circulation following intravenous administration (Gokhale *et al.*, 1997; Litzinger *et al.*, 1996; Saijo *et al.*, 1994) (Table 1). For example, the plasma elimination of DC-chol/DOPE:18mer PS oligonucleotide complexes from the circulation was so rapid as to be unmeasurable (Litzinger *et al.*, 1996). In another study (Saijo *et al.*, 1994), complexes formed between Lipofectin™ (commercially available vesicles composed of equimolar DOTMA and DOPE) and ISIS 3466 (20mer phosphorothioate oligonucleotide targeted against p120) were eliminated from the circulation at the same rate as free oligonucleotide. Similar observations have been made for complexes of cationic lipids and plasmids (Osaka *et al.*, 1997). Large lipid:oligonucleotide complexes rapidly accumulate to high levels in the lungs, primarily because they become trapped in pulmonary capillaries. Subsequent release of oligonucleotide and/or lipid from the lung leads to redistribution and accumulation to high levels in the liver (Litzinger *et al.*, 1996). The use of a molar excess of cationic lipid in liposomes and lipid:oligonucleotide complexes promotes extensive interactions with plasma proteins (Semple *et al.*, 1998), which are virtually all net negatively charged at physiological pH. This results in their rapid elimination from the circulation (Semple *et al.*, 1998) and, in addition, may actually inhibit transfection and delivery to target cells (Litzinger *et al.*, 1996; Zelphati *et al.*, 1998). Furthermore, cationic liposomes have been shown to induce plasma turbidity and clotting activity (Senior, 1991), activate complement (Baron *et al.* 1998; Plank *et al.*, 1996 and Figure 1), and induce liver toxicity (Semple *et al.*, unpublished observations). Amphipathic polyethylene glycol (PEG) has been used as a steric barrier to reduce interactions with plasma proteins and increase circulation times of liposomes. In the presence of excess cationic lipid, however, PEG-lipid is relatively ineffective at increasing circulation times as extensive interactions with plasma proteins still occur (Semple *et al.*, 1998).

The short circulation lifetimes described above have significantly hampered attempts to demonstrate biological activity for complexes *in vivo* following systemic administration. Of the three studies cited above in which cationic lipid:oligonucleotide complexes were administered intravenously, only one attempted to measure the antisense effect *in vivo*. Gokhale *et al.* (1997) injected i.v. a 15mer phosphorothioate anti raf-1 antisense complexed with PC/chol/DDAB (the type of PC used was not specified). The authors determined that the complex was eliminated with a halflife of 24 min, compared to less than 5 min for the free oligonucleotide (Table 1) and they reported variable effects on the downregulation of raf-1 in a human xenograft tumor model. This was interpreted as being due to inconsistent extravasation of lipid:oligonucleotide complexes into the tumor site. Consistent reduction of raf-1 levels were achieved only after direct intratumoral administration, but the effects on tumor progression were not reported (Gokhale *et al.*, 1997).

Other investigators employing cationic lipid:oligonucleotide complexes have avoided the problems associated with intravenous injection by using regional administration. This includes treatment of subcutaneous solid tumors by direct intratumoral injection (Marzo *et al.*, 1997; Sacco *et al.*, 1998) and intraperitoneal delivery to ascites tumors (Perlaky *et al.*, 1993; Saijo *et al.*, 1994). In the study by Perlaky *et al.* (1993), complexation of Lipofectin™ with ISIS 3466, a 20mer phosphorothioate against p120, substantially increased oligonucleotide potency in the peritoneal cavity. That is, decreasing the IC50 from > 10 mg/kg for the free ISIS 3466 to 0.26 mg/kg for the complex; however, *in vivo* protein or mRNA decreases were not reported (Perlaky *et al.*, 1993). It is likely that increased delivery of oligonucleotide to the tumor cells in the peritoneal cavity was achieved through an electrostatic interaction with the positively charged complexes. This is consistent with the observation that cationic lipid complexes of ISIS 3466 increased cellular accumulation of the oligonucleotide by 4- to 10-fold in an i.p./i.p. model (Saijo *et al.*, 1994). Sacco *et al.* (1998) reported that spontaneously arising mammary tumors in transgenic mice that were injected with complexes of DOSPER and a 13.5 Kb antisense plasmid against *neu* mRNA were significantly smaller than untreated contralateral tumors. However, these studies did not include control groups in which free antisense plasmid or DOSPER complexes with appropriate (i.e. sense, scrambled, inverse etc.) non-functional plasmids were also administered.

One of the more comprehensive studies concerning regional administration of cationic lipid:oligonucleotide complexes is that by Marzo *et al.* (1997). They administered a complex of DOTAP with a 15mer phosphorothioate oligonucleotide against TGFβ2 by intratumoral injection to subcutaneous AC29 murine mesotheliomas. Significant reductions in the rate of tumor growth were observed for lipid complexes in conjunction with the antisense sequence, but not for lipid-alone or lipid complexes with a scrambled sequence. Antitumor activity was reportedly associated with decreases in TGFβ2 mRNA levels relative to β-actin as well as lower proportions of proliferating cells (BrdUrd+) compared to tumors injected with DOTAP:scrambled oligonucleotide controls. Furthermore, the increase in antitumor activity observed for the DOTAP:antisense complex did not appear to be due to sequence specific immune stimulation (Krieg *et al.*, 1995), as the proportions of tumor-infiltrating lymphocytes, CD4+, CD8+ or NK cells were the same for tumors treated with antisense or scrambled control.

It is clear that more effort must be directed towards the rational design of lipid-based delivery vehicles for the specific purpose of achieving effective systemic delivery of antisense oligonucleotides. While neutral carriers display excellent pharmacokinetics and characteristics that enable accumulation at disease sites, they also have very low encapsulation efficiencies for charged antisense molecules. In contrast, cationic lipid:oligonucleotide complexes can provide high levels of encapsulation, but their size and charge severely impair their ability to distribute systemically after parenteral administration and subsequently extravasate to disease sites. As a result, these types of carriers have a limited utility for the treatment of systemic disease (Litzinger, 1997).

10.3 SYSTEMIC DELIVERY OF ANTISENSE DRUGS BY CONVENTIONAL LIPOSOMES

We have reviewed the evidence that liposome-based delivery systems can dramatically alter the pharmacokinetics and biodistribution of an encapsulated drug. If the carrier exhibits the necessary characteristics it will accumulate within tumors and at sites of inflammation or infection. This phenomenon of disease-site targeting is believed to play a major role in the enhanced efficacy observed for a variety of conventional drugs when formulated inside lipid vesicles (Bakker-Woudenberg *et al.*, 1992; Bally *et al.*, 1994), but what about antisense oligonucleotides? The study by Wielbo *et al.* (1996) has already been discussed in the previous section. They demonstrated that the potency of a phosphorothioate antisense targeted against angiotensinogen mRNA was increased when it was delivered i.v. in PC/chol liposomes. We have also observed enhanced potency of antisense drugs when they are encapsulated in PC/chol vesicles and our experience is discussed below.

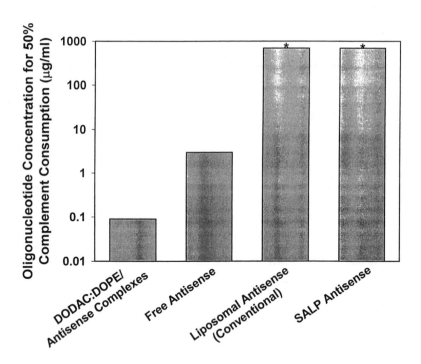

Figure 1. Complement activation by free antisense and various lipid-based delivery systems for antisense. The concentration of oligonucleotide required to consume 50% of complement in an *in vitro* assay is shown for free oligonucleotide (ISIS 3082), ISIS 3082 in DODAC:DOPE complexes, passively encapsulated in conventional 0.1 μm egg PC/chol liposomes or encapsulated in the SALP. Note (*) that values presented for both conventional and SALP formulations are minimum values; no complement consumption was observed for these carriers at the highest concentration tested (700 μg/ml).

To investigate the effects altered pharmacokinetics and biodistribution have on the biological activity of antisense drugs, we characterized an acute inflammation model to study the activity of anti murine ICAM-1 antisense. The model is based on a delayed-type hypersensitivity (DTH) response to cutaneous contact with the chemical irritant dinitrofluorobenzene (DNFB). Painting the ears of sensitized mice with a solution of DNFB triggers a cascade of pro-inflammatory events, resulting in a temporary remodeling of the local dermal vasculature that allows leakage of macromolecules and extravasation of cells into the peripheral tissue. The inflammatory response can generally be measured in terms of four basic endpoints: erythema (redness), edema (swelling), vascular leak (of proteins and small molecules), and cellular infiltration (Klimuk *et al.*, 1998a).

One of the primary inflammatory events initiated following epicutaneous exposure to DNFB is an increased expression of ICAM-1 on the luminal surface of local endothelial cells, keratinocytes, and antigen-presenting cells in the dermis. This protein assists in antigen presentation to T-lymphocytes and enhances leukocyte trafficking to the inflammation site by binding circulating leukocytes to the walls of local blood vessels (van de Stolpe and van der Saag, 1996). Diapedesis and migration of leukocytes into the extravascular tissue follow, and it is at this phase of the inflammatory process that lipid vesicles also extravasate (disease-site targeting). It has been shown that inhibiting over expression of ICAM-1 through antisense mechanisms or blocking the binding between ICAM-1 and its ligand, leukocyte function-associated antigen 1 (LFA-1), ameliorate inflammatory responses *in vivo* (Bennett *et al.*, 1997; Stepkowski *et al.*, 1994).

A 20mer phosphorothioate antisense (ISIS 3082) that hybridizes to the 3′ untranslated region of murine ICAM-1 mRNA with high affinity (Stepkowski *et al.*, 1994), and has demonstrated efficacy in several animal models (Bennett *et al.*, 1997; Stepkowski *et al.*, 1994), was passively encapsulated in 100 nm PC/chol vesicles. As expected, the effects of encapsulation on the behavior of the antisense *in vivo* were substantial. First, it was noted that the encapsulated phosphorothioate no longer activated human complement *in vitro* (Figure 1), a well known side effect of phosphorothioate backbone oligonucleotides (Henry *et al.*, 1997). These data are consistent with the liposomal membrane preventing exposure of the oligonucleotide to plasma proteins. In contrast, when the oligonucleotide was complexed with DODAC:DOPE, the ability of the formulation to activate complement was significantly increased (Figure 1). Second, the circulation half-life of the oligonucleotide was increased from just a few minutes to approximately 8 h, this in turn resulted in a significant accumulation of carrier and drug at the site of inflammation (Figure 2). Finally, the encapsulated antisense was significantly more potent than free drug when administered i.v. as a single bolus injection at the time inflammation was initiation. After 24 h, the ears of animals receiving empty vesicles alone, free ISIS 3082, an encapsulated scrambled control sequence (ISIS 8997) or an encapsulated non-binding control against human ICAM-1 (ISIS 2302) had doubled in thickness (Figure 3). In marked contrast, the inflammatory response in animals injected with encapsulated anti ICAM-1 antisense was greatly reduced, and similar to the response measured in the positive control group that received topical corticosteroid (Klimuk *et al.*, 1998b).

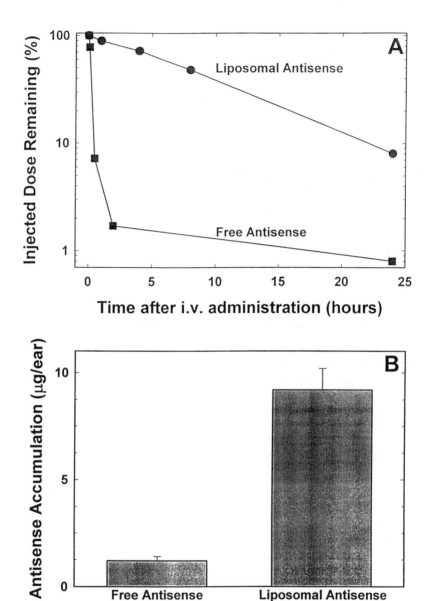

Figure 2. (A) Comparison of the pharmacokinetics of free and liposomal antisense oligonucleotides after intravenous injection. ISIS 3082 was injected into mice at a dose of 50 mg/kg as either free oligonucleotide or passively encapsulated in 0.1 μm egg PC/chol liposomes. The halflife of clearance of the oligonucleotide is increased from about 5 minutes for the free form to approximately 8 hours in the egg PC/chol liposomes. (B) Disease-site accumulation of oligonucleotide in inflamed ears at 24 hours after administration. Free ISIS 3082 or 3082 passively encapsulated in 0.1 μm egg PC/chol liposomes were administered to mice in the acute ear inflammation model.

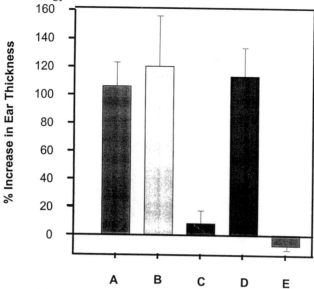

Figure 3. Summary of the *in vivo* efficacy of ISIS 3082 in an acute murine ear inflammation model. No efficacy ($\approx$ 100% increase in ear thickness) was observed for mice treated with empty egg PC/chol liposomes (A), free ISIS 3082 (B) or liposomes encapsulating a scrambled control sequences (D). In marked contrast, the efficacy of ISIS 3082 passively encapsulated in 0.1 μm egg PC/chol liposomes (C) was equivalent to that of the positive control, topical corticosteroid treatment (E). All oligonucleotides doses shown were 50 mg oligonucleotide/kg body weight.

10.4 STABILIZED ANTISENSE LIPID PARTICLES

In the previous section we highlight the potential advantages afforded to antisense drugs when they are delivered in an encapsulated form after intravenous injection. But conventional liposomes are not practical drug delivery vehicles for charged oligonucleotides because of their low encapsulation efficiency and poor fusogenic characteristics. As a result, we have developed a delivery system for anionic antisense drugs that incorporates the desirable features of both cationic lipid complexes and conventional liposomes (Semple *et al.*, manuscript in preparation). Referred to as Stabilized Antisense-Lipid Particles (SALP), an example is depicted schematically in Figure 4. The SALP has a flexible modular, or cassette, design in which components can be added, removed or replaced with alternatives according to the needs of the therapeutic agent and the biology of the disease being treated. Furthermore, the method by which the particles are assembled employs mild conditions and is straightforward to scale-up. This is important because the technology is to be developed for clinical applications and therefore must conform to both commercial and regulatory requirements.

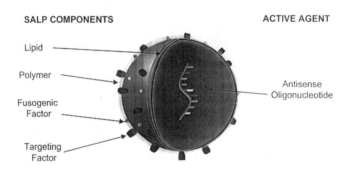

Figure 4. Schematic representation of the Stabilized Antisense Lipid Particle (SALP). The components include neutral structural lipids, a cationic lipid for electrostatic association with the oligonucleotide and a polymer-conjugated lipid. Other components can include targeting ligands and/or fusogenic factors.

Key physicochemical characteristics of SALP

SALP are designed specifically for intravenous injection and treatment of systemic disease. The core components of the formulation are: (i) an ionizable cationic lipid with a pKa between pH 5.0 and 6.5; (ii) a PEG-conjugated lipid, and; (iii) neutral structural lipids. Encapsulation and particle formation are achieved in a spontaneous self-assembly process that occurs when the components are mixed in an aqueous/ethanol solution. The pH is maintained below the pKa of the ionizable cationic lipid during the formulation step. This provides the necessary positive charge to drive an electrostatic interaction between lipid and oligonucleotide. The presence of the polymer-lipid conjugate controls particle aggregation caused by oligonucleotide-mediated cross-linking during self-assembly, but does not inhibit the association of antisense molecules with individual particles. The neutral lipids employed in the delivery system described here are PC and cholesterol, which help provide a stable bilayer for the final particle.

The process employed to generate SALP normally produces a population of vesicles with mean diameters that range from 150-200 nm. Further processing by extrusion (Semple *et al.*, manuscript in preparation) can produce more homogeneous populations with a narrower size distribution. Repeated passage through filters with a 100 nm pore size, for example, will produce SALP with average diameters in the

100-120 nm range. We routinely confirm vesicle size by quasi-elastic light scattering, although more labor-intensive methods such as freeze-fracture and cryo electron microscopy have been used. When vesicle sizing is complete, the solution pH is raised to pH 7.4-7.6. In this pH range, greater than 90% of the cationic lipid should be uncharged and unencapsulated oligonucleotide and ethanol are removed by standard techniques such as dialysis, tangential flow diafiltration, size exclusion chromatography (Figure 5) or ion exchange chromatography. However, the particle membrane is capable of maintaining a transbilayer pH gradient, therefore, the aqueous core of the SALP remains somewhat acidic. As a result the inside surface of the membrane is positively charged with antisense drug bound to the lipid/aqueous interface. So far the formulation technique appears to work for a variety of oligonucleotide chemistries and is apparently independent of nucleotide sequence, as long as the molecule being encapsulated is polyanionic (Semple *et al.*, manuscript in preparation).

The encapsulation efficiency is typically determined as the oligonucleotide/lipid ratio compared before and after gel filtration or DEAE-Sepharose column chromatography (Figure 5). For most oligonucleotides, efficiencies of 65-80% and final drug/lipid ratios of 0.15 to 0.2 (w/w) are readily achieved. In contrast, under the same conditions as those described in Figure 5, removal of the ionizable lipid or mixing of the antisense at pH > 7.4 typically results in encapsulation efficiencies of about 3% and a final drug/lipid ratio of 0.01 to 0.02 (w/w).

As with conventional liposomes, SALP completely protect the encapsulated oligonucleotide from interaction with plasma proteins, such that at physiological pH (surface charge neutral) they do not activate complement (Figure 1) and the antisense is protected from degradation by nucleases for at least 24 h (data not shown). In contrast, free oligonucleotide and oligonucleotide released from the carrier by detergent solubilization is completely degraded by S1 nuclease within 5 minutes.

Pharmacokinetic properties of SALP

SALP particles exhibit a net neutral surface charge, consequently *in vivo* they behave similarly to the conventional PC/chol vesicles described in Section 10.3. However, the circulation lifetime of the delivery vehicle can be modulated by the nature of the PEG-lipid component. In the particles described here, we employ a PEG molecule conjugated to the hydrophilic headgroup of a ceramide (Webb *et al.*, 1998b). The lipid anchors the PEG moiety to the SALP membrane. It is well documented that opsonization of liposomes *in vivo* is inhibited when the membrane surface is protected by a coating of PEG (Chonn *et al.*, 1992); consequently, the PEG-coated liposomes exhibit extended circulation times. We also observe this for SALP, however the effect is dependent on the length of the amide-linked hydrocarbon anchor. This determines the off-rate at which the PEG-ceramide molecule diffuses from the outer monolayer of the particle *in vivo*. PEG-ceramide containing a 20-carbon amide remains associated with the carrier for at least 24 h and exhibits extended circulation times. On the other hand, PEG-ceramides with 14-carbon amide chains have off-rates of just a few minutes *in vivo*, and particles formulated with this lipid exhibit half-lives similar to conventional liposomes (Figure 2A) of the same size (Semple *et al.*, manuscript in preparation). This demonstrates the flexibility of the SALP formulation; the optimum composition will likely depend upon the target mRNA and nature of the disease being treated.

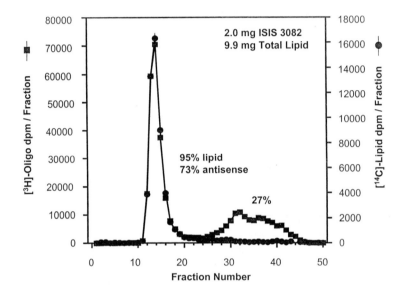

Figure 5.   Gel filtration chromatography of an SALP formulation of ISIS 3082. The 20mer phosphorothioate antisense oligonucleotide against murine ICAM-1 was encapsulated in the SALP carrier as described in the text. Gel filtration chromatography demonstrates that the 73% of the oligonucleotide is encapsulated in the SALP.

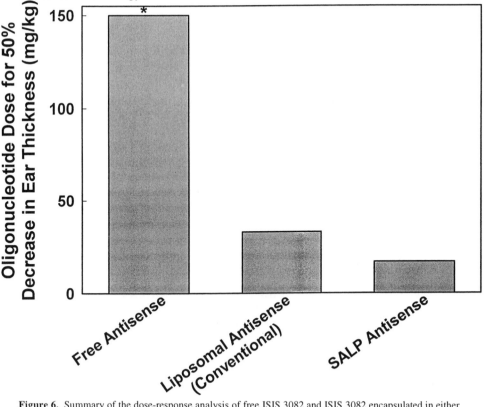

Figure 6. Summary of the dose-response analysis of free ISIS 3082 and ISIS 3082 encapsulated in either conventional (EPC/chol) liposomes or SALP. Efficacy is plotted as the interpolated oligonucleotide dose required to achieve a 50% inhibition of ear swelling. In this model, free ISIS 3082 had no detectable efficacy (*) up to the highest dose tested of 150 mg/kg.

Biological consequences in cancer and inflammation models

In Section 10.3, the enhanced efficacy of ISIS 3082 in an acute murine ear inflammation model was described. We have further examined the enhanced *in vivo* efficacy of SALP-encapsulated oligonucleotides in this inflammation model. The dose-response characteristics of these different formulations shows that free ISIS 3082 has no activity up to doses of 150 mg/kg. Significant increases in activity were observed when ISIS 3082 was encapsulated in either lipid-based delivery vehicle. That is, for ISIS 3082 in conventional 0.1 μm egg PC/chol liposomes, the dose for 50% inhibition of ear swelling was approximately 32 mg/kg (Figure 6). Superior *in vivo* efficacy was observed when ISIS 3082 was encapsulated in SALP, displaying a dose for 50% inhibition of swelling of 17 mg/kg (Figure 6).

10.5 SUMMARY AND CONCLUSIONS

Many of the *in vivo* studies of lipid-based delivery vehicles have employed formulations that were effective *in vitro*. However, this approach does not recognize the basic distinction between delivery to a monotype of cells in culture dishes and complexities of pharmacokinetics and biodistribution after parenteral administration. That is, while cationic liposome:oligonucleotide complexes can achieve high "encapsulation" efficiencies, their use *in vivo* is very limited due to short circulation lifetimes, minimal tumor site accumulation and only partial protection from nuclease degradation. Nonetheless, formulations based on cationic lipid:oligonucleotide complexes may have some utility for local administration purposes. In contrast, neutral carriers avoid the circulation lifetime/size/charge issues associated with cationic lipid:oligonucleotide complexes, but have low encapsulation efficiencies and are not commercially viable. The SALP carrier described here is a rationally designed for the *in vivo* delivery of anionic antisense oligonucleotides. It combines the high encapsulation efficiency characteristics of cationic liposome:oligonucleotide complexes with the longer circulation lifetimes associated with neutrally charged carriers. SALP enable passive targeting of antisense drugs to disease sites and accumulation in extravascular spaces that are not accessible to the larger cationic lipid:oligonucleotide complexes. Consequently, they confer significant increases in both the dose intensity and dose duration at disease sites.

We are currently evaluating the SALP carrier for antisense oligonucleotides for *in vivo* activity against a variety of oncology and inflammation targets. In addition, we are actively modifying the particles to include components that may further enhance intracellular delivery (i.e. pH-sensitive lipids, polymers, peptides, etc) and disease-site specific targeting. It is our expectation that these alterations will continue to improve the biological activity of antisense drugs. Other researchers have developed interesting formulations, but most have only been evaluated *in vitro*. These include (1) the minimum volume entrapment anionic system (Thierry and Dritschilo, 1992a; Thierry and Dritschilo, 1992b), (2) various targeted systems using heme (Takle *et al.*, 1997), antibodies (Leonetti *et al.*, 1990; Zelphati *et al.*, 1994a) and folate (Wang *et al.*, 1995); (3) neutral systems achieving high encapsulation efficiencies by rendering the oligonucleotides more lipophilic using conjugation to cholesterol (Oberhauser and Wagner, 1992) or alternative chemistries (Tari *et al.*, 1994, 1998); (4) pH-sensitive delivery systems (Couvreur *et al.*, 1997; De Oliveira *et al.*, 1997; Milhaud *et al.*, 1996); (5) combination of pH-sensitive and folate-targeted systems using polylysine-condensed oligonucleotides (Li and Huang, 1997, 1998), and; (6) combination of oligonucleotide conjugation to cholesterol and encapsulation in antibody-targeted immunoliposomes (Zelphati *et al.*, 1994b).

In summary, we have demonstrated that lipid-based delivery vehicles dramatically change the pharmacokinetics and biodistribution of antisense drugs, leading to significant accumulation of oligonucleotide at sites of disease which can result in enhanced efficacy.

10.6 REFERENCES

Allen TM, Hansen C, Rutledge J. Liposomes with prolonged circulation times: factors affecting uptake by reticuloendothelial and other tissues. *Biochim Biophys Acta* 1989;981:27-35

Aoki M, Morishita R, Higaki J, Moriguchi A, Kida I, Hayashi S, Matsushita H, Kaneda Y, Ogihara T. In vivo transfer efficiency of antisense oligonucleotides into the myocardium using HVJ-liposome method. *Biochem Biophys Res Com* 1997;231:540-45

Bakker-Woudenberg IA, Lokerse AF, Ten Kate MT, Storm G. Enhanced localization of liposomes with prolonged blood circulation time in infected lung tissue. *Biochim Biophys Acta* 1992;1138:318-26.

Bally MB, Masin D, Nayar R, Cullis PR, Mayer LD. Transfer of liposomal drug carriers from the blood to the peritoneal cavity of normal and ascitic tumor-bearing mice. *Canc Chemother Pharmacol* 1994;34:137-46

Bally MB, Nayar R, Masin D, Hope MJ, Cullis PR, Mayer LD. Liposomes with entrapped doxorubicin exhibit extended blood residence times. *Biochim Biophys Acta* 1990;1023:133-39

Barron LG, Meyer KB, Szoka FC Jr. Effects of complement depletion on the pharmacokinetics and gene delivery mediated by cationic lipid-DNA complexes. *Hum Gen Ther* 1998;9:315-23

Bennett CF, Kornbrust D, Henry S, Stecker K, Howard R, Cooper S, Dutson S, Hall W, Jacoby HI. An ICAM-1 antisense oligonucleotide prevents and reverses dextran sulfate sodium-induced colitis in mice. *J Pharmacol Exp Therapeut* 1997;280:988-1000

Bergan R, Connell Y, Fahmy B, Neckers L. Electroporation enhances c-myc antisense oligodeoxynucleotide efficacy. *Nucl Acid Res* 1993;21:3567-73

Boerman OC, Oyen WJG, Storm G, Corvo ML, van Bloois L, Van der Meer JWM, Corstens FHM. Tecnetium-99m labeled liposomes to image experimental arthritis. *An Rheumat Dis* 1997;56:369-73

Boman NL, Cullis PR, Mayer LD, Bally MB, Webb MS. Liposomal vincristine: The central role of drug retention in defining therapeutically optimized anticancer formulations. Woodle MC and Storm G eds. In: "Long circulating liposomes: old drugs, new therapeutics", pp. 29-49. 1997; Georgetown, Texas, Landes Bioscience

Boman NL, Masin D, Mayer LD, Cullis PR, Bally MB. Liposomal vincristine which exhibits increased drug retention and increased circulation longevity cures mice bearing P388 tumors. *Cancer Res* 1994;54:2830-33

Buhr CA, Wagner RW, Grant D, Froehler BC. Oligodeoxynucleotides containing C-7 propyne analogs of 7-deaza-2'-deoxyguanosine and 7-deaza-2'-deoxyadenosine. *Nucl Acid Res* 1996;24:2974-80

Chonn A, Cullis PR. Recent advances in liposomal drug-delivery systems. *Curr Opinion Biotechnol* 1995;6:698-708

Chonn A, Semple SC, Cullis PR. Association of blood proteins with large unilamellar liposomes in vivo. *J Biol Chem* 1992;267:18759-65

Clarenc JP, Lebleu B, Leonetti JP. Characterization of the nuclear binding sites of oligodeoxyribonucleotides and their analogs. *J Biol Chem* 1993;268:5600-04

Couvreur P, Fattal E, Malvy C, Dubernet C. pH-sensitive liposomes: an intelligent system for the delivery of antisense oligonucleotides. *J Lipid Res* 1997;7:1-18

Crooke ST. Delivery of oligonucleotides and polynucleotides. *J Drug Target* 1995;3:185-90

De Oliveira MC, Fattal E, Ropert C, Malvy C, Couvreur P. Delivery of antisense oligonucleotides by means of pH-sensitive liposomes. *J Controlled Rel* 1997;48:179-84

Fenlon CH, Cynamon MH. Comparative in vitro activities of ciprofloxacin and other 4-quinolones against Mycobacterium tuberculosis and Mycobacterium intracellulare. *Antimicrob Agents Chemother* 1986;29:386-88

Flanagan WM, Wagner RW. Potent and selective gene inhibition using antisense oligodeoxynucleotides. *Mol Cel Biochem* 1997;172:213-25

Gay JD, DeYoung DR, Roberts GD. In vitro activities of norfloxacin and ciprofloxacin against Mycobacterium tuberculosis, M. avium complex, M. chelonei, M. fortuitum, and M. kansasii. *Antimicrob Agents Chemother* 1984;26:94-96

Giles RV, Spiller DG, Grzybowski J, Clark RE, Nicklin P, Tidd DM. Selecting optimal oligonucleotide composition for maximal antisense effect following streptolysin O-mediated delivery into human leukaemia cells. *Nucl Acid Res* 1998;26:1567-75

Gokhale PC, Soldatenkov V, Wang F-H, Rahman A, Dritschilo A, Kasid U. Antisense raf oligodeoxyribonucleotide is protected by liposomal encapsulation and inhibits Raf-1 protein expression in vitro and in vivo: implications for gene therapy of radioresistant cancer. *Gen Ther* 1997;4:1289-99

Hangai M, Tanihara H, Honda Y, Kaneda Y. In vivo delivery of phosphorothioate oligonucleotides into murine retina. *Arch Opthalmol* 1998;116:342-48

Henry SP, Giclas PC, Leeds J, Pangburn M, Auletta C, Levin AA, Kornbrust DJ. Activation of the alternative pathway of complement by a phosphorothioate oligonucleotide: potential mechanism of action. *J Pharmacol Exp Therapeut* 1997;281:810-16

Hope MJ, Bally MB, Webb G, Cullis PR. Production of large unilamellar vesicles by a rapid extrusion procedure. Characterization of size distribution, trapped volume and ability to maintain a membrane potential. *Biochim Biophys Acta* 1985;812:55-56

Hope MJ, Mui B, Ansell S, Ahkong QF. Cationic lipids, phosphatidylethanolamine and the intracellular delivery of polymeric, nucleic acid-based drugs. *Mol Membran Biol* 1998;15:1-14

Klimuk SK, Semple SC, Scherrer P, Hope MJ. Contact hypersensitivity: a simple model for the characterization of disease site targeting by liposomes. 1998a Submitted

Klimuk SK, Semple SC, Nahirney PN, Bennett CF, Scherrer P, Hope MJ. Enhanced anti-inflammatory activity of liposomal ICAM-1 antisense oligonucleotide. 1998b Submitted

Krieg AM, Yi A-K, Matson S, Waldschmidt TJ, Bishop GA, Teasdale R, Koretzky GA, Klinman DM. CpG motifs in bacterial DNA trigger direct B-cell activation. *Nature* 1995;74:46-49

Leonetti J-P, Machy P, Degols G, Lebleu B, Leserman L. Antibody-targeted liposomes containing oligodeoxyribonucleotides complementary to viral RNA selectively inhibit viral replication. *Proc Nat Acad Sci USA* 1990;7:448-51

Li S, Huang L. Targeted delivery of antisense oligodeoxynucleotides by LPDII. *J Liposome Res* 1997;7:3-75

Li S, Huang L. Targeted delivery of antisense oligodeoxynucleotides formulated in a novel lipidic vector. *J Liposome Res* 1998;8:239-50

Litzinger DC. Limitations of cationic liposomes for antisense oligonucleotide delivery in vivo. *J Liposome Res* 1997;7:51-61

Litzinger DC, Brown JM, Wala I, Kaufman SA, Van GY, Farrell CL, Collins D. Fate of cationic liposomes and their complex with oligonucleotide in vivo. *Biochim Biophys Acta* 1996;1281:139-49

Liu D, Liu F, Song YK. Recognition and clearance of liposomes containing phosphatidylserine are mediated by serum opsonin. *Biochim Biophys Acta* 1995;1235:140-46

Marzo AL, Fitzpatrick DR, Robinson BWS, Scott B. Antisense oligonucleotides specific for transforming growth factor b2 inhibit the growth of malignant mesothelioma both in vitro and in vivo. *Canc Res* 1997;57:3200-07

Mayer LD, Bally MB, Cullis PR. Strategies for optimizing liposomal doxorubicin. *J Liposome Res* 1990a;1:463-80

Mayer LD, Bally MB, Cullis PR, Wilson SL, Emerman JT. Comparison of free and liposome encapsulated doxorubicin tumor drug uptake and antitumor efficacy in the SC115 murine mammary tumor. *Canc Lett* 1990b;53:183-90

Mayer LD, Bally MB, Loughrey H, Masin D, Cullis PR. Liposomal vincristine preparations which exhibit decreased drug toxicity and increased activity against murine L1210 and P388 tumors. *Canc Res* 1990c;50:575-79

Mayer LD, Nayar R, Thies RL, Boman NL, Cullis PR, Bally MB. Identification of vesicle properties that enhance the antitumour activity of liposomal vincristine against murine L1210 leukemia. *Canc Chemother Pharmacol* 1993;33:17-24

Mayer LD, Tai LCL, Ko DSC, Masin D, Ginsberg RS, Cullis PR, Bally MB. Influence of vesicle size, lipid composition, and drug-to-lipid ratio on the biological activity of liposomal doxorubicin in mice. *Canc Res* 1989;49:5922-30

Milhaud PG, Bongartz JP, Lebleu B, Philippot JR. pH-sensitive liposomes and antisense oligonucleotide delivery. *Drug Deliv* 1996;3:67-73

Morishita R, Gibbons GH, Kaneda Y, Ogihara T, Dzau VJ. Pharmacokinetics of antisense oligodeoxyribonucleotides (cyclin B1 and CDC 2 kinase) in the vessel wall in vivo: enhanced therapeutic utility for restenosis by HVJ-liposome delivery. *Gene* 1994a;149:13-19

Morishita R, Gibbons GH, Ellison KE, Nakajima M, von der Leyen H, Zhang L, Kaneda Y, Ogihara T, Dzau VJ. Intimal hyperplasia after vascular injury is inhibited by antisense cdk2 kinase oligonucleotides. *J Clin Invest* 1994b;93:1458-64

Moulds C, Lewis JG, Froehler BC, Grant D, Huang T, Milligan JF, Matteucci MD, Wagner RW. Site and mechanism of antisense inhibition by C-5 propyne oligonucleotides. *ACS Biochem* 1995;34:5044-53

Oberhauser B, Wagner E.. Effective incorporation of 2'-O-methyl-oligoribonucleotides into liposomes and enhanced cell association through modification with thiocholesterol. *Nucl Acid Res* 1995;20:533-38

Osaka G, Carey K, Cuthbertson A, Godowski P, Patapoff T, Ryan A, Gadek T, Mordenti J. Pharmacokinetics, tissue distribution, and expression efficiency of plasmid [33P]DNA following intravenous administration of DNA/cationic lipid complexes in mice: use of a novel radionuclide approach. *J Pharmaceut Sci* 1997;85:612-18

Parr MJ, Masin D, Cullis PR, Bally MB. Accumulation of liposomal lipid and encapsulated doxorubicin in murine lewis lung carcinoma: the lack of beneficial effects by coating liposomes with poly(ethylene glycol). *J Pharmacol Exp Therapeut* 1997;280:1319-27

Perlaky L, Saijo Y, Busch RK, Bennett CF, Mirabelli CK, Crooke ST, Busch H. Growth inhibition of human tumor cell lines by antisense oligonucleotides designed to inhibit p120 expression. *Anti-Canc Drug Design* 1993;8;3-14

Plank C, Mechtler K, Szoka FC Jr, Wagner E. Activation of the complement system by synthetic DNA complexes: a potential barrier for intravenous gene delivery. *Hum Gen Ther* 1996;7:1437-46

Sacco MG, Barbieri O, Piccini D, Noviello E, Zoppé M, Zucchi I, Frattini A, Villa A, Vezzoni P. In vitro and in vivo antisense-mediated growth inhibition of a mammary adenocarcinoma from MMTV-neu transgenic mice. *Gen Ther* 1998;5:388-93

Saijo Y, Perlaky L, Wang H, Busch H. Pharmacokinetics, tissue distribution, and stability of antisense oligodeoxynucleotide phosphorothioate ISIS 3466 in mice. *Oncol Res* 1994;6:243-49

Sanders CC, Sanders WE Jr, Goering RV. Overview of preclinical studies with ciprofloxacin. *Am J Med* 1987;82:2-11

Semple SC, Chonn A, Cullis PR. Interactions of liposomes and lipid-based carrier systems with blood proteins: relation to clearance behavior in vivo. *Adv Drug Deliv Rev* 1998;32:3-18

Senior JH, Trimble KR, Maskiewicz R. Interaction of positively-charged liposomes with blood: implications for their application in vivo. *Biochim Biophys Acta* 1991;1070:173-79

Spiller DG, Giles RV, Grzybowski J, Tidd DM, Clark RE. Improving the intracellular delivery and molecular efficacy of antisense oligonucleotides in chronic myeloid leukemia cells: a comparison of streptolysin-O permeabilization, electroporation, and lipophilic conjugation. *Blood* 1998;91:4738-46

Stepkowski SM, Tu Y, Condon TP, Bennett CF. Blocking of heart allograft rejection by intercellular adhesion molecule-1 antisense oligonucleotides alone or in combination with other immunosuppressive modalities. *J Immunol* 1994;153:5336-46

Takle GB, Thierry AR, Flynn SM, Peng B, White L, Devonish W, Galbraith RA, Goldberg AR, George ST. Delivery of oligoribonucleotides to human hepatoma cells using cationic lipid particles conjugated to ferric protoporphyrin IX (heme). *Antisense & Nucl Acid Drug Dev* 1997;7:177-85

Tardi PG, Boman NL, Cullis PR. Liposomal doxorubicin. *J Drug Target* 1996;4:129-40

Tari AM, Tucker SD, Deisseroth A, Lopez-Berestein G. Liposomal delivery of methylphosphonate antisense oligodeoxynucleotides in chronic myelogenous leukemia. *Blood* 1994;84:601-07

Tari AM, Lopez-Berestein G. Oligonucleotide therapy for hematological malignancies. *J Liposome Res* 1997;7:19-30

Tari AM, Stephens C, Rosenblum M, Lopez-Berestein G. Pharmacokinetics, tissue distribution, and safety of P-ethoxy oligonucleotides incorporated in liposomes. *J Liposome Res* 1998;8:251-64

Thierry AR, Dritschilo A. Intracellular availability of unmodified, phosphorothioated and liposomally encapsulated oligodeoxynucleotides for antisense activity. *Nucl Acid Res* 1992a;20:5691-98

Thierry AR, Dritschilo A. Liposomal delivery of antisense oligodeoxynucleotides. Application to the inhibition of the multidrug resistance in cancer cells. *Ann NY Acad Sci* 1992b;660:300-02

Tomita N, Morishita R, Higaki J, Aoki M, Nakamura Y, Mikami H, Fukamizu A, Murakami K, Kaneda Y, Ogihara T. Transient decrease in high blood pressure by in vivo transfer of antisense oligonucleotides against rat angiotensinogen. *Hypertension* 1995;26:131-36

van de Stolpe A, van der Saag PT. Intercellular adhesion molecule-1. *J Mol Med* 1996;74:13-33

Vlassov VV, Balakireva LA, Yakubov LA. Transport of oligonucleotides across natural and model membranes. *Biochim Biophys Acta* 1994;1197:95-108

Wang S, Lee RJ, Cauchon G, Gorenstein DG, Low PS. Delivery of antisense oligodeoxyribonucleotides against the human epidermal growth factor receptor into cultured KB cells with liposomes conjugated to folate via polyethylene glycol. *Proc Nat Acad Sci USA* 1995;92:3318-22

Webb MS, Harasym TO, Masin D, Bally MB, Mayer LD. Sphingomyelin-cholesterol liposomes significantly enhance the pharmacokinetic and therapeutic properties of vincristine in murine and human tumour models. *Br J Canc* 1995;72:896-904

Webb MS, Boman NL, Wiseman DJ, Saxon D, Sutton K, Wong KF, Logan P, Hope MJ. Antibacterial efficacy against an in vivo Salmonella typhimurium infection model and pharmacokinetics of a liposomal ciprofloxacin formulation. *Antimicrob Agent ChemoTher* 1998a;42:45-52

Webb MS, Saxon D, Wong FMP, Lim HJ, Wang Z, Bally MB, Choi LSL, Cullis PR, Mayer LD. Comparison of different hydrophobic anchors conjugated to poly(ethylene glycol): effects on the pharmacokinetics of liposomal vincristine. *Biochim Biophys Acta* 1998b; in press

Wielbo D, Simon A, Phillips MI, Toffolo S. Inhibition of hypertension by peripheral administration of antisense oligonucleotides. *Hypertension* 1996;28:147-51

Yamada K, Moriguchi A, Morishita R, Aoki M, Nakamura Y, Mikami H, Oshima T, Ninomiya M, Kaneda Y, Higaki J, Ogihara T. Efficient oligonucleotide delivery using the HVJ-liposome method in the central nervous system. *Am J Physiol* 1996;271:R1212-R1220

Zalipsky S, Brandeis E, Newman MS, Woodle MC. Long circulating, cationic liposomes containing amino-PEG-phosphatidylethanolamine. *FEBS Lett* 1994;353:71-74

Zelphati O, Imbach J-L, Signoret N, Zon G, Rayner B, Leserman L. Antisense oligonucleotides in solution or encapsulated in immunoliposomes inhibit replication of HIV-1 by several different mechanisms. *Nucl Acid Res* 1994a;22:4307-14

Zelphati O, Wagner E, Leserman L. Synthesis and anti-HIV activity of thiocholesteryl-coupled phosphodiester antisense oligonucleotides incorporated into immunoliposomes. *Antivir Res* 1994b;25:13-25

Zelphati O, Uyechi LS, Barron LG, Szoka FC Jr. Effect of serum components on the physico-chemical properties of cationic lipid/oligonucleotide complexes and on their interactions with cells. *Biochim Biophys Acta* 1998;1390:119-33.

Carriers for delivery of antisense drugs

TABLE 1: Summary of the published *in vivo* evaluations of antisense oligonucleotides in lipid-based delivery systems

Liposome Composition	Size (nm)	Gene target	Encap. Effic. (%)	Admin. Route; T1/2α (h)	Notes	Ref.
Cationic lipid:antisense complexes						
DOTAP:15mer PS oligonucleotide	Large[1]	TGF-β2	100[1]	Intratumoral; N/A	Delayed AC29 s.c. tumor growth; decreased TGF β2 mRNA	Marzo *et al.*, 1997
DOSPER:13.5 Kb antisense plasmid	Large[1]	*Neu*	100[1]	Intratumoral; N/A	Decreased tumor growth; decreased *neu* mRNA & protein	Sacco *et al.*, 1998
PC/chol/DDAB (3.2/1.6/1);15mer PO antisense[2]	1-10 μm	Raf-1	> 90	i.v.; 0.41	Decreased raf-1 protein levels, no tumor growth data	Gokhale *et al.*, 1997
DOTMA/DOPE:20mer PS oligonucleotide (Lipofectin)	Large[1]	p120	100[1]	i.v.; 0.23	Free oligo = Lipofectin:oligo. i.v. Lipofectin increased uptake i.p.	Saijo *et al.*, 1994
DOTMA/DOPE:20mer PS oligonucleotide (Lipofectin)	Large[1]	p120	100[1]	i.p.; N/A	Efficacy of Lipofectin:oligo >> free oligonucleotide	Perlaky *et al.*, 1993
DC-chol/DOPE (1/1);18mer PS oligonucleotide	1.8 μm	None	100[1]	i.v.; Very short[3]	Rapid (5 min) accumulation in lungs; long term accumul. in liver	Litzinger *et al.*, 1996
Net neutral liposomes						
DOPC; 18mer P-ethoxy oligonucleotide	≤ 900 nm	Various	> 95	i.v.; 0.11	Increased circulation halflife	Tari *et al.*, 1998
PC/cholesterol (8/2); 18mer PS oligonucleotide	≈ 100 nm[1]	Angioten-sinogen	< 10[1]	intra-arterial; ND	Decreased plasma AngII levels, decreased blood pressures	Wielbo *et al.*, 1996

[1] not determined directly; estimated from preparation procedure.
[2] oligodeoxyribonucleotide; diester linkages with terminal phosphorothioate at 5' and 3' ends.
[3] not readily quantifiable; 80% of injected dose was present in the lungs at 5 min post-administratio

11 Designing a Clinical Antisense Study

Finbarr E Cotter and Dean Fennell

LRF Molecular Haematology Unit
Institute of Child Health
London, UK

11.1 INTRODUCTION

Greater understanding of the molecular basis of many diseases has led to exploration of "gene silencing" strategies of which antisense oligonucleotides have been in the forefront. Developments in this field have now entered the clinical arena with variable results. However, it is becoming clear that some of these molecules may eventually become medicines. Considerable challenges associated with robust therapeutic studies of antisense oligonucleotides remain. Historically drug development has involved an empirical approach.Therapeutic effect has been shown often independently of any specific knowledge concening underlying pharmacodynamics. As long ago as two hundred years it was known that bark of the white willow, *salix alba* contained a glycoside, salicin with potent anti-inflammatory activity. In the mid-19th century, Kolbe published methods for the synthesis of salicylic acid and aspirin appeared as a medicine in 1899. However, it was not until 1971 that John Vane and colleagues in London elucidated the mode of action as being inhibition of prostaglandin synthesis. This empirical path of has been the main paradigm for discovery of lead drugs, and the scientific study of therapeutic effects since the mid-1940s has not required *a priori* demonstration of mode of action at the molecular level. Antisense oligonucleotides pharmacology, however, has evolved in a reverse direction. The molecular genetics of disease have implicated expression of critical genes as the crucial factor in the disease process. New therapy is based on designing a drug that has a specific known action and then applying it to the disease process. Antisense oligonucleotides represent one such class of drugs and as such portray a new reductionist approach to drug design which in turn will require novel approaches to their clinical evaluation and eventual introduction into everyday medicine.

11.2 THE PURPOSE OF CLINICAL ANTISENSE OLIGONUCLEOTIDE STUDIES

Once shown to have efficacy *in vitro* and in animal models (subject to demonstration of reasonable safety) a candidate antisense oligonucleotide should be administered to humans in order to observe
· Therapeutic efficacy
· Toxicology profile
· Pharmacokinetics.

It is the demonstration of efficacy which will ultimately determine whether or not an antisense oligonucleotide will become a medicine. The most important response metameter in the clinical setting is therapeutic effect measured by established clinical-pathological correlates, but should also include a demonstration that the protein product of the targeted gene is reduced by administration of the antisense oligonucleotide. Although robust proof of antisense pharmacodynamics must be demonstrated in pre-clinical experiments, it may still be desirable to confirm this in the human setting. There are two important reasons for this:

Semantics

Antisense oligonucleotides are rationally designed and are unique in being named after their putative mechanism of action rather than their clinical effect (eg. anti-inflammatory compared with anti-depressant). As a consequence, it is logical that continued and appropriate use of such a description is dependent on showing some correlate of antisense action in the therapeutic setting. Antisense oligonucleotide drugs are expensive to develop and substantiation of effect may be essential in early clinical studies to establish validity as truly gene specific drugs. For example, clinical responses shown to be associated with no protein downregulation would strongly suggest a mode of action independent of inhibitory effects on gene expression.

Access to technology

In some disease conditions, it should be possible to directly measure the effects of antisense oligonucleotides. For example, Webb (Webb et al., 1997) in a phase I relapsed lymphoma trial of BCL-2 specific phosphorothioate antisense oligonucleotide (G3139), obtained tissue by fine needle aspiration and bone marrow aspiration. Downregulation of BCL-2 was measured using flow cytometry. It may not always be possible to obtain disease cells due to lack of safe accessability.

Because the specificity (and putative clinical efficacy) of a candidate antisense oligonucleotide is predicted from its sequence information, chemically, antisense oligonucleotides are a remarkably homogenous group of drugs. As such, provided good pre-clinical data is available confirming sequence specificity of action it may only be a necessity for the earliest drug trials to pursue rigorous clinical evidence of protein downregulation. It will then have been shown that as a generic group of antisense oligonucleotides do indeed translate effectively from animal models to the human setting.

Toxicology

Therapeutic index (TI) measures the horizontal distance between the dose-response relationships for desirable and undesirable antisense oligonucleotide responses. If this distance is small, the risk of toxicity increases dramatically with incremental dose rendering the agent more hazardous. Clinical experience to date with phosphorothioate antisense oligonucleotides has shown these agents to have a relatively high therapeutic index, suggesting their potential clinical usefulness.

Pharmacokinetics

Inevitably, antisense oligonucleotide efficacy is dependent upon the drug reaching its target in sufficient quantity and for sufficient duration. The administration, distribution, metabolism and excretion of antisense oligonucleotide will influence this considerably and must therefore be studied to enable optimal scheduling to be achieved.

Antisense oligonucleotide efficacy versus antisense potency

The objective of antisense oligonucleotide therapeutic evaluation must be the same as for any other drug, that is to establish predictions of treatment outcome with a degree of certainty. This requires the determination of an effective dose range by characterisation of a dose-response relationship. antisense oligonucleotide efficacy, the maximum effect which can be achieved, is of great significance being the major determinant of clinical usefulness. At the molecular level, this may relate to the choice of RNA target sequence chosen for the design of the antisense oligonucleotide. This contrasts with therapeutic potency, defined as the efficacy in relation to the amount of drug. This in turn may be related to pharmacokinetic factors influenced by for example backbone chemistry. Potency is considered to be of less clinical significance compared with efficacy provided, the difference is not too large and the TI is sufficiently large (Lawrence et al., 1992).

Optimal responses for two antisense oligonucleotides of equal efficacy but different potency can be achieved simply by increasing dose. If two such antisense oligonucleotides X and Y are to be compared, the ratio of equally effective doses measures relative therapeutic potency i.e. the horizontal distance between respective parallel/ sigmoidal dose response curves.

11.3 CLINICAL ANTISENSE OLIGONUCLEOTIDE STUDIES

Design considerations

Initial studies of antisense oligonucleotide conducted in small groups such as phase I trials to primarily determine the safety of the antisense oligonucleotide molecule (10-50) and phase II (50-300) trials are aimed at crudely defining efficacy, toxicology and pharmacokinetic profiles. Phase I studies for antisense oligonucleotide essentially commence therapy at a dose (at least least 1/10th) well below the equivalent in the animal models that shows 10% mortality (the LD10) and increases the dose in increments with a number of patients treated at each dose (eg. following European Organisation for Research and Treatment of Cancer [EORTC, 1994] dose escalation recomendations). Toxicity is examined (graded by common criteria from 1 [mild] to 4 [severe] criteria) as well as a maximum tolerated dose (MTD). In essence when an unacceptable toxicity is observed (grade 3 or 4) at a particular dose the antisense oligonucleotide is stopped. Further subjects are treated at the same dose and if at least 50% of patients suffer similar grades of toxicity it, is deemed the MTD.

It is also essential to examine the effect of downregulating the target gene in normal tissue and the toxicity of the backbone chemistry in the Phase I study. The value of such trials depends highly on aspects of their general design which should consider the following:

Study objectives

The objective of ethically answering a precisely framed question should be at the centre of any well designed clinical investigation (Bradford Hill, 1977). In the case of antisense oligonucleotides, perhaps the simplest question for investigation that can be framed for the small clinical investigation may be; "Does antisense oligonucleotide X produce a desirable clinical effect Z with the minimum of side effects." In the small clinical investigation (phase I, below 30 subjects) it may be desirable but not essential to confirm an antisense mode of action by demonstration of target protein downregulation, although as discussed above this may not always be possible.

Subjects and entry criteria

The implication of a specific overexpressed target gene in the disease process is a prime entry criteria for an antisense oligonucleotide that downregulates that gene. The subjects need to have confirmed the presence of that gene.

In addition, there will be generally at least one of three indications for investigating the therapeutic effects of an antisense oligonucleotide. These are Failure of conventional treatment (for example refractory lymphoma or leukaemia), considerable potential for improvement upon existing therapies (such as inflammatory bowel disease) or the absence of available treatment, such as in many orphan diseases (uncommon conditions for which no prior drug development has existed).

Patients exhibiting the specific catagory of disease rather than volunteers, will almost invariably represent the subject pool of antisense oligonucleotide efficacy investigations. This however is not as essential for studies focusing on pharmacokinetic or toxocological properties. Phase I studies of efficacy in patients may employ measurements of therapeutic effect in relation to pre-existing disease i.e. historical controls. Such within-patient studies have been described as almost always unacceptable, owing to variability in both diagnosis and spontaneous fluctuations in severity (especially in chronic stable conditions). This problem was summarised by the eminent pharmacologist J.H. Gaddum who stated, "Patients may recover in spite of drugs or because of them." There are however set criteria (Bailar et al., 1984) which if adhered to may limit erroneous conclusions being drawn from such studies.

Ethics issues regarding antisense oligonucleotide controls

Proof of sequence specificity requires the use of at least one control (ideally both scrambled and sense controls) to validate an antisense mode of action and must be demonstrated in an *in vivo* animal model system prior to human trial. Proof of antisense action in humans presents an ethical challenge. If patients rather than volunteers are chosen as subjects for therapeutic investigation (as is likely to be the case), it is the ethical duty of the doctors charged with their care to offer the best treatment available to them (the so called principle of *non-maleficence*). Indications leading to the selection of a particular antisense oligonucleotide treatment will inevitably be based upon a detailed understanding of the molecular pathology. Thus, the doctor will be in posession of prior evidence that there is likely to be a difference between control and treatment. However in patients it is unethical to treat with scrambled or null controls. This problem must, however, be balanced against the risk of drawing false negative conclusions from an uncontrolled study which is in itself inherently unethical. Clearly, in diseases in which there are commonly spontaneous fluctuations (chronic diseases such as inflammatory bowel disease) such controls may be essential. Although presenting a design constraint, it is indeed possible to design and execute carefully planned studies aimed at answering a precisely framed question.

Antisense oligonucleotide administration and duration of treatment

The kinetics of antisense oligonucleotide action at the molecular level dictate the speed at which responses are able to occur. Existing target protein degradation must take place once gene expression is arrested. This will require firstly an appropriate duration of treatment determined by the half-lives of the mRNA and target protein. For example, BCL-2 has been shown to have a half life of approximately 10-14 hours (Reed et al., 1993) requiring at least 3 days treatment *in vitro* and *in vivo* (Cotter et al., 1994) to observe effective downregulation. Secondly, steady state plasma levels are required to prevent intermittent re-synthesis of protein, capable of abrograting cellular antisense responsiveness

Optimal dosing schedules should be based upon existing animal model studies. Webb (Webb et al., 1997) employed an out-patient administration protocol (after 24 hours in-patient observation) using subcutaneous infusion over a 2 week period, based on previous animal studies (Cotter et al., 1996).

Monitoring therapeutic effect

Measurement of clinical antisense oligonucleotide effects

Antisense oligonucleotides are unique in that they are designed to produce specific and potentially measurable effects at the level of gene expression in addition to associated clinical endpoints. Measurement of these response metameters is therefore desirable at least in small scale clinical investigations.

Protein downregulation as a biological endpoint

Tissue is necessary for the measurement of protein level which may be either assayed by flow cytometry or western analysis (RT-PCR is less desirable as it is relatively difficult to obtain a degree of quantification). Certain diseases (particularly haematological) permit the measurement of target protein more readily than others. For example, in the phase I lymphoma trial (Webb et al., 1997) protein was measured in fine needle aspirates of lymph nodes and bone marrow aspirates followed by flow cytometric analysis.

Relability of protein reduction measurements

Repeated measurement of protein levels is essential for gaining any quantitative understanding of central tendency (average) and dispersion (variance) providing information on experimental reliability. This is can be achieved using flow cytometry provided great care is exercised to ensure normalisation relative to a positive control (which represents 100% protein or σmax) and negative control (representing 0%, σmin), and is measured under identical assay conditions so as to minimise systemic bias. Normalisation of measured median fluorescence intensity (MFI) corresponding to the level of target protein (σ %) is calculated from the equation (see Fig.1) where σ is the MFI for antisense oligonucleotide treated sample, σmin negative control, and σmax positive control (e.g. derived as a historical control). Specificity may be determined by comparison of target protein levels to that of a control protein which is not to be downregulated by antisense oligonucleotide.

$$\sigma\% = \frac{\sigma - \sigma min}{\sigma max - \sigma min} \times 100$$

Fig. 1. Normalisation equation for expression of protein

Measurement of early molecular responses

In some instances it may be possible to measure highly specific biological responses resulting from treatment with antisense oligonucleotide. In the case of BCL-2 antisense treatment for lymphoma, apoptosis activity can be quantified using annexin V binding at single cell resolution via flow cytometry or caspase-3 activity (both these factors should be activated if BCL-2 is downregulated). Such measurements can provide corroborative data supporting a predicted mode of action and demonstrating effectiveness in the clinical setting. Such approaches do however require controls which may raise ethical problems.

Clinical endpoints

Demonstration of clinical improvement (or not) in the patient measured by clinical-pathological criteria is the primary purpose of the clinical investigation. Scoring systems should be available for this purpose. For example, tumour bidimensional product (BDP) may be used to assess treatment outcome; Miller et al (1981, World Health Organisation criteria) have defined 4 outcome groups on the basis of BDP measurement. Thus,

·Stable disease : No change in BDP
·Progressive disease : > 25% increase in BDP
·Partial response : > 50% decrease in BDP
·Complete response: disappearance of tumour

Proportions can be determined for treatment outcomes and inferences made regarding potential antisense oligonucleotide efficacy.

Surrogate endpoints

It is of considerable use to have early and predictive markers of effective therapy. The progress of antisense oligonucleotide treatment may be monitored by assessment of clinical parameters which are related to the main primary pathology per se, correlating with the changes in the severity of disease. For example serum lactate dehydrogenase may be followed as a biological marker for lymphoma monitoring. Such endpoints are useful to provide early information regarding possible clinical effectiveness, albeit with varying degrees of certainty, as compared with true clinical endpoints which may take several weeks to manifest.

11.4 THERAPEUTIC EVALUATION OF ANTISENSE OLIGONUCLEOTIDES

In the development of antisense oligonucleotide therapy, once initial clinical safety has been demonstrated in preliminary small scale studies, it is then essential for more robust investigations to be conducted so as to examine efficacy. Therapeutic evaluation involving carefully controlled studies should be able to address the following questions. The randomised control trial (Cochrane, 1972) is the established goal standard for clincal therapeutic evaluation.

A number of questions may be addressed. Is antisense oligonucleotide X with conventional treatment Y more potent than conventional treatment Y alone?·An example of this is the combination of BCL-2 antisense oligonucleotide and chemotherapy. If BCL-2, which mediates chemotherapy resistance, is downregulated are the effects of chemotherapy enhanced. Is antisense oligonucleotide X more or less potent than antisense oligonucleotide Y for the same target gene A? ·Is antisense oligonucleotide X more or less potent than antisense oligonucleotide Y for target genes A versus B? ·Does the combination of antisense oligonucleotide Y+Z (genes A and B) exhibit greater efficacy than antisense oligonucleotide X alone (gene A or B) ? However, before all these questions are designed into a phase II study, it is pivotal that downregulation of the target molecule by antisense oligonucleotide is demonstrated in a phase I study. A failure to show this negates the design and evaluation of further studies.

11.5 STATISTICAL CONSIDERATIONS

Randomisation

Systemic bias must be eliminated between treatment groups if a measured differences in clinical outcome are to be attributed to antisense oligonucleotide with any scientific validity. Random allocation of patients to achieve equivalent groups with significant differences in variables such as age, duration of disease, previous treatments and severity of disease remains the gold standard. In the case of antisense oligonucleotide evaluation, the possibility of a patient being randomised to receive control therapy (e.g. an existing treatment) which is potentially inferior to antisense oligonucleotide treatment or some combination, presents an ethical dilemma. Randomisation of patients to receive both conventional and test treatments, however, provides a general design strategy which can obviate this problem.

Cross-over designs

This design approach in its simplest form incorporates two treatment periods separated by a washout period. Patients are randomised to one of two groups which receive the two treatments in one and reverse order respectively.

One caveat to this approach is that persistent pharmacological activity of the active antisense oligonucleotide may lead to *carry-over* effects measured in a subsequent control period. Care must therefore be taken to minimise these effects in order to avoid erroneous conclusions being reached, and this may require care in determining the appropriate washout period to allow for carry-over effects to diminish. As a general example, X may be a conventional single agent chemotherapy treatment A, and Y a combination of antisense oligonucleotide, B (in this case an adjuvant) and A. For groups I and II of equal sizes n1 and n2 respectively, the differences in measured response for group I, $d_I = (j1 - j2)_I$ and group II, $d_{II} = (j1 - j2)_{II}$, should be equal if there is no treatment effect (null hypothesis). This can be tested statistically by using a standard 2-sample t-test on n_1+n_2 degrees of freedom, where the parameter $\sigma d2$ is pooled within-groups estimate of variance of d. This design allows quantification of carry-over effect and estimation of treatment magnitude can be made using this simple trial design.

Blinding

Subjective bias on the part of physican and patient who are aware of having been treated with antisense oligonucleotide may be very strong, particularly in view of the psychological influence of what may be perceived as a new and potentially effective drug. For this reason, random allocation should ideally incorporate doubling blinding. In some cases blinding of a patient may not be feasible of ethical.

Power of an antisense oligonucleotide trial

Observed differences, d in therapeutic outcome may on occasion differ from true differences D, owing to sampling variations in clinical responsiveness. The probability, b of failing to detect a difference because d < D which is not significantly different from zero is termed a type II error (i.e. make a false negative inference). This probability is determined by the experimental design, in particular, the number of subjects employed in the clinical trial which can be chosen by the investigator to minimise the likelihood of false negative conclusions. The power of a study is defined as the probability (1-b) of detecting a difference, i.e. avoiding a type II error. For a specified power, the required number of subjects can be estimated provided pilot information is available regarding the variance of measured clinical responses.

Sequential analysis

In the therapeutic setting, patients may not all be treated simultaneously, but in a temporally dispersed fashion. The size of the sample will ultimately be determined by the observed clinical responses and the precision attached to e.g in the cross-over design. The study may then be stopped at some point where a degree of precision can be assigned to differences in therapeutic effect. In this way, a stopping rule is required to be defined in planning. Regarding antisense oligonucleotide evaluation this may be a more favorable approach ethically in comparisons with conventional treatment where such a sequential approach should be mandatory, so as to avoid proceeding with a trial beyond a point where the antisense oligonucleotide may be seen to be producing a serious difference in clinical outcome. This general approach was originated by Wald (Wald, 1947).

11.6 SUMMARY

Antisense oligonucleotides represent a new class of pharmacological agent. It should provide individual downregulation of a single protein type, which may considerably reduce non-specific toxicity. However, non-sequence related toxicity can occur due to backbone chemistry and needs to be rigorously determined in order to understand the potential application and possible modifications of the sequence specific therapeutic molecules.

Antisense oligonucleotides differ in their relative lack of toxicity, which may be problematic when trying to determine the MTD as this could be well beyond the therapeutic dose. As such we could have to evaluate the need for continued dose escalation well beyond the dose required for theraputic benefit, in Phase I studies, in the future, particularly as increased efficacy and reduced toxicity are obtained with improved backbone chemistry and improved cellular uptake of antisense oligonucleotide. However, this may bring greater sequence specific toxicity's when genes are silenced in normal tissues where it is not beneficial. It is therefore essential for antisense oligonucleotides in clinical trials, that a broad evaluation of toxicity is undertaken, as well as clear demonstration of specific gene downregulation. In addition rigorous Phase I/II and III criteria should be followed.

11.7 REFERENCES

Bradford Hill A. *Principles of Medical Statistics*. London: Hodder and Stoughton, 1977

Bailar JC, Louis TA, Savori PW, Polansky M. Studies without internal controls. *N Engl J Med* 1984;311:156-62

Cochrane AL. The history of the measurement of ill health. *Int J Epidemiol* 1972;1:89-92

Cotter FE, Johnson P, Hall P, Pocock C, Mahdi NAl, Cowell JK, Morgan G. Antisense oligonucleotides suppress B-cell lymphoma growth in a SCID-hu mouse model. *Oncogene* 1994;9:3049-55

Cotter FE, Corbo M, Raynaud F, et al. Bcl-2 antisense therapy in lymphoma: In vivo and in vitro mechanisms, efficacy, pharmacokinetics and toxicity studies. *Ann Oncol* 1996;7:32

European Organization for Research and Treatment of Cancer (EORTC). *A practical guide to EORTC studies*. Place: EORTC 1994

Kitada S, Takayama S, De Riel K, Tanaka S, Reed JC. Reversal of Chemoresistance of lymphoma cells by antisense mediated reduction of bcl-2 gene expression. *Antisense Res Dev* 1994;4:71-79

Lawrence DR, Benett PN. *Clinical Pharmacology*. Churcill Livingstone, London UK 1992

Miller AB, Hoogstraten B, Staquet M, et al. Reporting results of cancer treatment. *Cancer* 1981;47:207-14

National Cancer Institute: *Guidelines for reporting of adverse drug reactions*: division of cancer treatment. Bethesda, MD:National Cancer Institute, 1988

Wald A. *Sequential analysis*. Wiley, New York, USA. 1947

Webb A, Cunningham D, Cotter F, Clarke PA, di Stefano F, Ross P, Corbo M, Dziewanoska. Bcl-2 antisense therapy in patients with non-Hodgkin lymphoma. *Lancet* 1997;349:112-37

INDEX